Russian Tanks and Armored Vehicles
1946-to the Present

Koch

Russian Tanks and Armored Vehicles

1946-to the Present

An Illustrated Reference

Fred Koch

Schiffer Military History
Atglen, PA

NORTH EASTERN LIBRARY SERVICE
AREA LIBRARY, DEMESNE AVENUE
BALLYMENA, Co. ANTRIM BT43 7BG

NUMBER 308703
CLASS 623.74750947

Book Design by Ian Robertson.

Copyright © 1999 by Koch.
Library of Congress Catalog Number: 99-62848

All rights reserved. No part of this work may be reproduced or used in any forms or by any means – graphic, electronic or mechanical, including photocopying or information storage and retrieval systems – without written permission from the copyright holder.

Printed in China.
ISBN: 0-7643-0914-5

We are interested in hearing from authors with book ideas on related topics.

Published by Schiffer Publishing Ltd.	In Europe, Schiffer books are distributed by:
4880 Lower Valley Road	Bushwood Books
Atglen, PA 19310	6 Marksbury Road
Phone: (610) 593-1777	Kew Gardens
FAX: (610) 593-2002	Surrey TW9 4JF
E-mail: Schifferbk@aol.com.	England
Visit our web site at: www.schifferbooks.com	Phone: 44 (0)181 392-8585
Please write for a free catalog.	FAX: 44 (0)181 392-9876
This book may be purchased from the publisher.	E-mail: Bushwd@aol.com.
Please include $3.95 postage.	
Try your bookstore first.	Try your bookstore first.

Contents

	Foreword	6
Chapter 1:	The Development of Armored Troops	7
Chapter 2:	The Russian Armored Vehicles of World War II and their Improvements in the Fifties and Sixties	11
Chapter 3:	The Path to the Modern Battle Tank	48
Chapter 4:	The Modern Russian Battle Tank of the Guard Tank Units - the T-80 Medium Tank	62
Chapter 5:	The Self-propelled Artillery of the USSR	65
Chapter 6:	Anti-aircraft Tanks and Mobile Anti-aircraft Rockets	72
Chapter 7:	Mobile Multipurpose Launchers - in Russian, "Battle Machines"	82
Chapter 8:	The Mobile Rocket Carriers of the Soviet Army	87
Chapter 9:	Tank Destroyers - The Tank-destroyer Guided Missiles on Motorized Launches	94
Chapter 10:	The Armored Reconnaissance Vehicles	98
Chapter 11:	Armored Transporters and Combat Vehicles of the Infantry	106
Chapter 12:	Special Armored Vehicles of the Soviet Army, its Allies and the Armies that Evolved after the Dissolving of the USSR	120
Chapter 13:	Recovery Tanks, Technical Service Vehicles and Tracked Barrier-crossing Technology	130
Chapter 14:	Special Towing Tractors on Tracked Chassis	142
Chapter 15:	Armored Vehicles of Air-landed Units	150
	Bibliography	158
	Schematics Gallery	159

Foreword

In the last decades, countless publications about the development of armored vehicles, weapons carriers and special tracked vehicles in Russia since the end of World War II have appeared. Every newly "revealed" vehicle has immediately towed a series of articles and reports behind it.

Suspicions and speculations about the increase in their combat value as well as their use by the Russian troops have often appeared accompanied by only a few photos. Every new detail, whether it is only the modified attachment of a chest, immediately led to a new designation. As a rule, rearmaments and modernizations caused problems. Remarkably, the Russian military publications were available to most authors of books about the armored vehicles of the Soviet army. Presumably, nobody trusted these publications. The Russian mentality, that reduces everything to the absolutely necessary, with an internationally incomparable degree of standardization, resulted in not only simple technical solutions but also very simple designations of the Soviet weapons of war.

This book is meant to offer, for the first time, as broad a spectrum as possible of the tanks and special vehicles used by the Russian troops. In the process, only Russian developments and licensed copies were considered. For all the illustrations it was assured that the vehicles shown did not differ from the original production other than in national and tactical markings and crew members. Modifications or improvements that were undertaken by other countries have been deliberately omitted. On the other hand, though, numerous projects and prototypes have been included, even if they were not supplied to the Russian armed forces. All the designations used correspond to those of the Russian forces, though some have been augmented with the NATO codes already familiar from previous publications.

At this point I would like to thank all those who provided support in the preparation of this book. In particular, I thank my wife, Anett, for the translation of many newspaper articles, Mr. Gert Herr for the creation of numerous diagrams, sketches and drawings, Mr. Hans-Jürgen Janaczeck and Mr. Andreas Gryscheck for the gathering of information and photographic material, Mr. Klaus Koch for his years of action with "hidden camera" in times of strict secrecy, and the staff of the Military History Museum of Dresden, who have assisted me in so many ways.

Dresden, February 1999

1

The Development of Armored Troops

The victory of the Red Army over Germany in World War II was utilized by the leadership of the USSR, especially in postwar years, as proof of the superiority of Soviet technology. Along with countless reports of the "heroic deeds" of the Russian soldiers in World War II, every article that appeared in the international press was used to promote a positive evaluation of the superiority of Russian war technology. This was interesting only in its simplicity. Among others, there quickly arose the myth that the T-34 was the best tank in World War II. In the end, though, only the design of the T-34 gained international recognition.

As early as 1943, a commission of the American test center in Aberdeen had come to the conclusion that the Russian T-34 represented a good tank design in its essential solutions and was suitable for mass production. In particular, several outstanding qualities were emphasized, such as the low streamlined shape, the simplicity of design, and the low ground pressure.

The T-34 attained the high point of its praise in a German publication in 1954. In it the author wrote: "From its first action on it was clear that this tank exceeded all other types in terms of firepower, armor and speed. Its shape was ideal and became a model for numerous succeeding types. Good off-road capability in mud and snow resulted from wide tracks. Very favorable specific attainments."[1]

Independent of the surely existing pride in the successful compromise among firepower, mobility and armor protection that was attained in the T-34, the Soviet military command recognized that the superior numbers of the tank on the battlefield had been decisive in attaining victory in World War II. The memoirs of numerous Russian generals particularly indicated this fact.

In May 1945, according to official statistics, 11,365,000 men belonged to the Red Army. By 1948 its strength had been reduced to 2,874,000 men.[2] This applied to all units of the Red Army except those of the armored and mechanized divisions. Their numbers, and the numbers of personnel involved in them, were still considerably heightened. When the war ended, there were supposed to be, in all, 25 armored and 13 mechanized divisions in existence. The rifle divisions also had a mixed armored and self-propelled gun regiment with two self-propelled gun batteries and five armored

Very few tanks have gained such a reputation in their history as the T-34. This is the last version produced during the war, the T-34-85 medium tank (1944 model), seen during combat in Berlin at the end of April 1945.

1. Mostovenko, W. D., Panzer gestern und heute, Berlin 1961, p. 156.

companies assigned to them. Along with the striven-for full motorization with tracked tractors and motor vehicles, the armored troop carriers in particular had become the infantry's most important means of transport.

Essential changes took place in the phase of demobilization in the leadership of the Red Army. Along with numerous units and military schools, corps, army and fleet commands were disbanded. The system of military districts was organized anew. Of the 33 military districts in existence at the end of 1945, there were only 21 in October 1946. On September 4, 1945, the State Defense Committee and the Headquarters of the Troop Commander were disbanded, as well. The command of all parts of the fighting forces was taken over by the newly created Ministry of Defense.

Around 1950 the Soviet armored regiment consisted of three tank battalions with 31 tanks or self-propelled guns each, plus one motorized rifle battalion. Every tank battalion was composed of three tank companies. Every company, consisting of three platoons, had a total of ten tanks or self-propelled guns. As a rule, the tank regiment was to have had a total of 65 medium tanks and 21 medium self-propelled guns on hand. A similar principle was applied to the motorized rifle units. Thus, the motorized rifle platoon likewise had three armored troop carriers. The mechanized divisions consisted of three mechanized regiments, one tank regiment with medium tanks, one heavy tank regiment with a total of 23 heavy tanks and 42 heavy self-propelled guns, an artillery regiment equipped with 122 mm howitzers, and the security units.[3] In 1954 a mechanized division was to be equipped with a total of 208 tanks and 63 self-propelled guns.

In 1948 a Russian rifle division had 1,488 motor vehicles, towing tractors and armored troop carriers. At the end of 1944 there had been only 419 vehicles. The rifle divisions had thus gained much in mobility and were armed nearly equally to the existing mechanized units by being equipped with tanks and self-propelled guns.

Although the question of the operative significance of tanks was raised anew with the introduction of stronger antitank guns, antitank guided missiles and antitank weapons attacking from the air, the tank remained the most important offensive weapon in the Soviet military doctrine.

The commander of the Soviet armored troops, Marshal Pavel Rotmistrov, formulated the following five points for the superiority of the tank as opposed to other types of weapons:

1. During an attack, the tank has the ability to close in on the enemy so closely that he can then destroy him with his heavy weapons.
2. Only the tank is capable of carrying on a constant attack while in motion and thus eliminating enemy nests of resistance that have escaped the artillery preparation. It succeeds by using fire and mobility.
3. Only the tank with its tank gun and machine guns is able to knock out all obstacles for the advancing infantry during the attack and at the same time fight against tanks and artillery.
4. Only the tank, thanks to its armor, can intervene in combat without having to fear machine-gun or artillery fire, automatic or manual weapons. In fact, it can fight them down by itself. By clever action it can even come out the victor.
5. Only the tank, because of its motor and tracks, can attack at high speed and destroy the enemy before the enemy is able to defend himself.[4]

By 1952 the strength of the tank and rifle units of the Soviet Army was to have reached a total of 57 tank, 39 mechanized and 55 rifle divisions. For 1953, 64 tank, 36 mechanized and 52 rifle divisions were acknowledged.[5]

With the introduction of rocket weapons in the weaponry of various armies, the significance of armored troops increased, and not only in the USSR. In May 1955 there appeared in the journal *Armor*: "The mobility of the tank, its firepower and striking power in massed action, and the protection that its armor affords against shock waves, light rays and initial rays make the tank an ideal weapon for use in atomic war."

At the end of the fifties, the newly established strategic rocket troops of the USSR were created as a further striking force. On July 15, 1946, the first unit armed with rockets was established by adapting it from a guard regiment. It was meant primarily to provide a noteworthy increase in the firepower and striking power of the ground forces. The rocket troops were regarded as the main means of carrying out tactical and operative combat tasks. But the Soviet military doctrine also followed the concept that only the tanks were in a position to be able to penetrate into the heart of the enemy forces after the use of nuclear weapons. Thus at the same time the significance of the antitank artillery increased. Along with the increase in penetrating power and the introduction of antitank guided missiles, the development concentrated particularly on the creation of armored means of towing, transport and carrying, which also had to be capable of being ready for action in the areas contaminated by nuclear weapons. In order to be able to guarantee a close cooperation between the armored troops and the motorized rifle units, the importance of the armored troop carriers as a means of infantry transport was affirmed. The anti-aircraft artillery and the anti-aircraft rocket troops were also given greater mobility and armor protection, in order to assure the attacking troops of constant cover from the air. The necessary fire-control systems were also installed in armored vehicles. At the beginning of the sixties, the air-landed troops of the Soviet Army received, in addition to new antitank and anti-aircraft weapons, modern information and command technology, and the first self-propelled mounts. The engineer troops were equipped with a series of special technical devices, most of which were installed on the chassis of their armored vehicles. These included road-building machines, mine-locating and -removing vehicles, plus bridge and pontoon carriers. The technical troops received armored repair and towing vehicles. The chemical units were equipped with armored vehicles for radiation and chemical detection.

By 1957, the rifle and mechanized divisions had become the motorized rifle divisions. As a rule, they were composed of three motorized rifle regiments, one tank regiment, and the special and

---

2. Skorobogatkin, K. F., Die Streitkräfte der UdSSR, Berlin 1974, p. 599.
3. Andronikov, I. G., Die roten Panzer, München 1963, p. 205.
4. Garforth, R., How Russia makes war, London 1954, no page number.
5. Ogorkiewicz, R. M., Armour, London 1960, no page number.

Chapter 1: The Development of Armored Troops

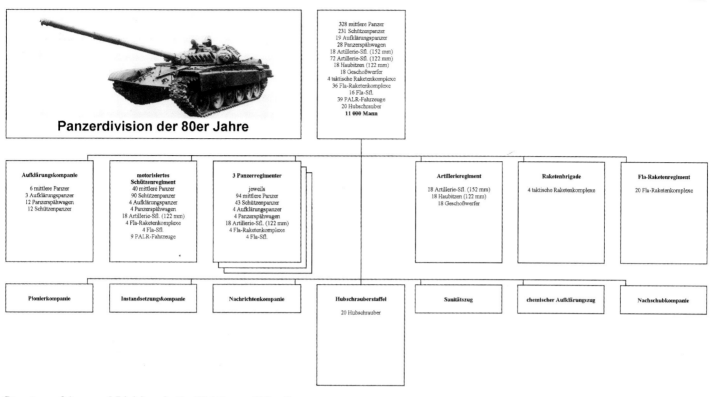

Structure of Armored Divisions in the Eighties and Nineties

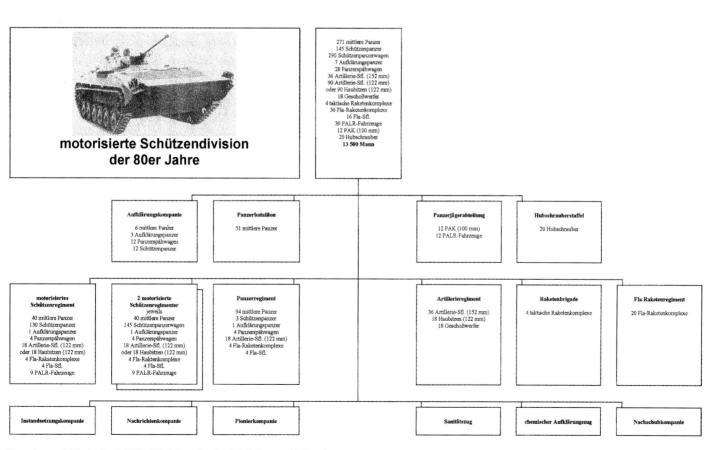

Structure of Motorized Rifle Divisions in the Eighties and Nineties

security troops. The armored divisions, likewise reorganized by 1957, included two regiments, each with 95 medium tanks, a heavy tank regiment with 95 heavy tanks, a motorized rifle regiment, an artillery regiment, and special and security units, in all some 11,000 men and some 285 tanks. Just two years before, the number of members of the Soviet Army had risen to 5,763,000 men.

For the year of 1967, the number of armored divisions in the Soviet Army were listed as 43 and the motorized rifle divisions as 90.

In the motorized rifle divisions, there were some 210 medium tanks on hand at this time. The self-propelled gun mounts had, for the most part, been mustered out step by step by that time. In place of them, the reconnanssaice companies of the armored and motorized rifle divisions were gradually being supplied with light amphibian tanks.

At the beginning of the seventies, a rifle division had sixteen times as many tanks, 37 times as many armored troop carriers, thirteen times as many automatic weapons and five times as many means of intelligence as it had had in 1939. The degree of motorization had risen from 3 HP to 30 HP per man, and the weight of a gun and grenade-launcher salvo from 1,700 kg to 53,000 kg.[6]

At the beginning of the eighties, the transition was made to equip the heavy tank regiments with medium tanks, thus turning them into "normal" tank regiments. But in 1991 there were still some armored divisions of the so-called shock armies that had heavy tank regiments.

By 1991 the numbers of armored units, and thus of tanks, increased steadily. During the Vienna Disarmament Conference of 1991, information about the tanks in the Soviet Army was published for the first time. According to it, in the area "from the Atlantic to the Ural Zone" there were over 20,500 battle tanks, of which 7,000 were to be scrapped by 1994.[7]

In connection with the Vienna disarmament talks, the Soviet military leadership saw itself compelled for the first time to depart from its doctrine of concentrated action with quantitative superiority. Through extensive modernization measures, a definite increase in superiority was to be achieved. Under the pressure of increasing economic and domestic political problems, the variety of Soviet tank model types in particular was to be reduced. An essential influence on the number of models to be mustered out was surely the experience that had been gained with Soviet tanks in Afghanistan, the Gulf War, the Caucasus region and the former Yugoslavia.

---

6. Schröder, S., 60 Jahre Sowjetarmee, Berlin 1981, p. 190.
7. Flotho, H., Entwicklung der sowjetischen Kampfpanzer nach Wien, Frankfurt am Main 1991, p. 762.

2

The Russian Armored Vehicles of World War II and their Improvements in the Fifties

After the evaluation of war experiences in the realm of tank development, the emphasis was placed on increasing firepower without having to limit mobility. In addition, a low silhouette, amphibious ability or a deep wading ability of at least four meters were required for the armored vehicles. Armored vehicles were also to be built for the air-landed units operating in the enemy's hinterlands. This was a requirement that had been repeated again and again ever since the thirties.

Although work had been done since 1944 on a successor model to the medium tank of World War II, for many years the armored units still consisted of tanks from the last war. Only the light tanks appeared to have been removed from the armaments completely. The T-70 light tank (1942 model) had, in fact, been manufactured only until 1944. Until the introduction of the PT-76 light amphibian tank (1951 model) in the spring of 1952, there were no light tanks in the Soviet Army's fighting forces. The first postwar suggestion for a light amphibian tank originated in 1950 and took shape on the chassis of the T-70 or SU-76, with the designation K-90. But it was only used for testing and never went into production.

Among the almost countless postwar developments is this K-? light tank, produced in the early sixties and intended for the airborne troops.

K-90 Light Amphibian Tank
Legkiy plavajushiy tank K-90 (obr. 1950 g.)

Made in: USSR (tank factory in Volgograd)
Used in: USSR
Developed: 1950
Manufactured: 1950
Crew: 3 men
Fighting weight: 10 tons
Armament: 1 76.2 mm tank gun (40 rounds)
                1 7.62 mm machine gun (1,000 rounds)
Powerplant: 6-cylinder M-2 Diesel engine (140 HP/103 kW)
Top speed: 43 kph on road, 9.6 kph in water
Range: 250 km

All armored divisions were at first equipped only with medium and heavy tanks. The medium tanks were then responsible for "tank vs. tank" fighting and penetrating into the depths of the enemy defenses. The heavy tanks were to open the attack and break down the enemy's defenses to allow his defensive positions to be penetrated. The available self-propelled guns were to take part in the artillery's preparation, remove nests of opposition during the

By the end of World War II, the Red Army had a great number of various light tanks. The T-70 light tank (1942 model), which the Russian armament industry produced until 1944, was the last light tank in series production into the fifties.

For many years the Soviet Army and its allies used World War II tanks. Until the fifties, these vehicles equipped the armored and mechanized rifle divisions. This picture shows the joint division drills of the Soviet forces in Germany and the Nationale Volksarmee of East Germany on August 17-23, 1957. The tank column is led by a T-34 medium tank (1942 model) dating from 1942.

The most important Russian tank of World War II was the T-34 medium tank (1943 model). It too was used by the Soviet Union and its allies for many years after the war. This picture shows one of the few surviving vehicles of the former East German Army, with the author driving.

attack, and take part in antitank action. In addition, they were to cover the heavy tanks.

Among the tanks and self-propelled gun mounts of World War II there were the T-34 medium tank (1943 model) and T-34-85 (1944 model), the IS-2 heavy tank (1943 model), IS-2 (modernized) (1943 model) and IS-3 (1945 model), the SU-76m (1943 model) light self-propelled gun mounts (in Russian, "self-propelled artillery set-ups"), the SU-85 (1943 model) and SU-100 (1944 model) medium self-propelled mounts, and the ISU-122 (1943 model), ISU-122s (1943 model) and ISU-152 (1943 model) heavy gun mounts. In addition, the reconnaissance units utilized the BA-64 (1942 model) armored car produced during the war.

The T-34 medium tank (1943 model), equipped with the 76.2 mm tank gun (Russian: Tankowaja puschka), was available only in the reserve camps of the Soviet Army and among its allies in the first postwar years. At times, though, these tanks were also used for

This maneuver photo shows that the Soviet Army also used its T-34 tanks until the fifties. A T-34 medium tank (1943 model) is shown along with a BTR-152 armored transporter (1950 model).

Most of the tanks built during the war ended like this T-34 (1943 model), as sitting targets on one of the many troop training camps of the Soviet Army and its allies.

Chapter 2: The Russian Armored Vehicles of World War II

The T-34 medium tank (1943 model), reequipped with camouflaged headlight and other lights for use in the East German Army. For museum purposes, this vehicle was given the red star of the Russian Army.

training. Modernizing them was never considered. The situation was different for the exported T-34 tanks. For example, the nineteen T-34 (1943 model) tanks that were turned over to the East German tank readiness and tank school were equipped with radio sets and an aiming telescope in the turret, which were not included when the tanks were delivered. With the building up of the armed forces in the countries occupied by the Soviet Army, more and more T-34 (1943 model) tanks came to the armies of the Eastern Block states. The number of these tanks in East Germany had already reached 100 by the latter half of 1952.

In the mid-fifties, though, they were mustered out and rebuilt. In the last two war years, a few tanks had already been rebuilt as towing and salvage vehicles. Numerous T-34 (1943 model) tanks withdrawn from active service were also rebuilt into tracked towing tractors after the war.

The production of the T-34-85 (1944 model) medium tank, on the other hand, continued until 1947. The modernized T-34-85 (1947 model) was even built until June 1964, though as of 1955-56 the production was taken over by the CSSR and Poland. The improved T-34-85 (1947 model) had a new ventilation system for the fighting compartment. Armored ventilation shafts were installed in the turret in the center of the front and behind the entry hatches. The turret and the tank gun could also be aimed electrically. The improved 10 RT radio apparatus was then installed, as well as a gearbox with five forward speeds and one reverse. Now the quality of workmanship was also considerably sturdier. Among other things, the welded seams were made more neatly. In all, more than 12,000 T-34-85 (1944 and 1947 models) were produced, and they were utilized in 46 countries of the world. After the armored regiments of the Warsaw Pact nations were supplied with modern battle tanks, many of these T-34-85 (1944 and 1947 models) were delivered to Third World nations, and some of them remain in use to this day. Small numbers of these tanks were also rebuilt as recovery tanks, or ended as "hard targets" on tank firing ranges. Large numbers were also in the reserve camps of the Warsaw Pact nations until 1989-90.

Medium Tank T-34-85, T-34-85M
Sredniy tank T-34-85 (obr. 1947 g.), T-34-85M

Made in: USSR (tank factories in Kharkov, Leningrad, Nishni Tagil, Omsk), Czechoslovakia (Pilsen), Poland (Katowitze)
Used in: USSR, Afghanistan, Albania, Algeria, Angola, Austria, Bangladesh, Bulgaria, Cambodia, China, Congo, Cuba, Cyprus, CSSR, East Germany, Egypt, Equatorial Guinea, Ethiopia, Finland, Guinea, Guinea-Bissau, Hungary, Iraq, Israel, Laos, Libya, Mali, Mongolia, Mozambique, Nicaragua, North Korea, North Yemen, Poland, Romania, Somalia, South Yemen, Sudan, Syria, Tanzania, Togo, Uganda, United Arab Emirates, Vietnam, Yugoslavia, Zambia, Zimbabwe.
Developed: 1947.
Manufactured: 1947-1956 (USSR), 1951-1964 (CSSR, Poland)
Crew: 5 men.
Fighting Weight: 32.0 tons
Overall length: 8150 mm
Hull length: 6070 mm
Width: 2950 mm
Height: 2720 mm
Armament: 1 85 mm ZIS-S-53 L/54.6 tank gun (1944 model) (60-80 rounds);
2 7.62 mm DTM machine guns (2750 rounds)
Powerplant: 12-cylinder W-2-34 Diesel engine (500 HP / 368 kW)
Top speed: 55 kph
Range: 300 km

Until replaced by more modern types, the T-34 medium tank (1944 model) remained the most important battle tank of the Soviet Army and its allies. This one is seen at a tank driving school on November 27, 1958.

Production of the T-34 (1944 model) continued until 1947. Its off-road capability under rough conditions particularly won international respect.

After the war, the lower bow panel ahead of the driver was reinforced on most T-34-85 tanks, for this formed a vertical target surface when the tank came out of low spots.

As mentioned above, work on a new battle tank was begun as early as mid-1944. The reason was the necessity of having to install a tank gun with an even larger caliber than 85 mm in medium tanks. Since a large-caliber gun would result in a greater weight for the turret, first a way had to be found to locate the center of gravity as near as possible to the center of the tank's hull, without having to depart from the space division into driver's area, fighting compartment and engine room. In addition, the rear mounting of the gearbox, the side shafts and drive wheels was regarded as dependable. This was to be accepted by all the international tank manufacturers in the fifties. The only way the turret could be located nearer to the center consisted of shortening the engine room, which was finally

Many of these T-34s also became targets at one of the many troop training camps in the end.

A T-34-85 (1944 model) with the typical covers for fighting-compartment ventilation, as made during the war.

Chapter 2: The Russian Armored Vehicles of World War II

In 1947 production of the modernized T-34 began. Along with the modified covers for the fighting-compartment ventilation, the T-34-85 medium tank (1947 model) was also fitted with a five-speed gearbox. This one is seen in attack training in 1963.

Production of the T-34-85 (1947 model) continued until June 1964. As of 1955-56 the tank factories in the CSSR and Poland took over its production.

A T-34-85 (1947 model) at a tank driving school.

The 12-cylinder, 500 HP W-2-34 Diesel engine, which gave the T-34 and its derivatives better than average performance.

For lack of suitable recovery vehicles, many T-34 tank chassis were rebuilt into armored towing tractors after the tanks were mustered out. This T-34-T medium armored towing tractor of a NVA unit is seen during the "Quartette" maneuvers in October 1963.

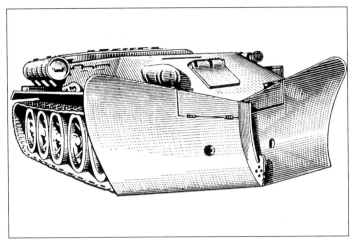

A T-34-T medium armored towing tractor with STU dozer (1952 model).

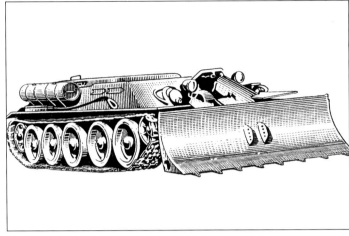

To remove obstacles and build field positions, several T-34-T medium armored towing tractors were fitted with the BTU grading dozer (1958 model).

A T-34-T medium armored towing tractor (1957 model) with a special body for spare parts and special tools.

For field repairs, several T-34-T medium armored towing tractors were fitted with a small crane.

A T-34-T medium armored towing tractor with a different type of crane.

Chapter 2: The Russian Armored Vehicles of World War II

To be able to lift heavy burdens, several tank hulls were rebuilt into the SPK-5 medium field repair cranes. Able to lift up to five tons, they could carry out large repair jobs in the field.

Night firing drill for an armored troop equipped with T-44 or T-44S (1966 model) medium tanks. The T-44M had the engine of the T-54 and T-55 medium tanks and infra-red equipment. From 1966 on the T-44S (1966 model) has been modernized with stabilizers for the tank gun.

accomplished by mounting the motor transversely. The space thus gained farther forward allowed the installation of the entry hatch on the top of the hull and the use of a homogeneous front armor. By 1946 the testing of the hull was essentially finished. After three Russian tank factories had begun work on the T-44 (1944 model) in the same year, at the end of 1946 and the beginning of 1947 a few armored guard regiments were equipped with these tanks.

Since the development of a new tank gun with a rotating turret was not yet finished, the vehicles were fitted with the T-34-85 turret. In the second half of 1946 the Soviet tank designers finally succeeded in developing an appropriate tank turret, which could hold a 100 mm tank gun and could be installed on the hull used for the T-44 (1944 model) tank. This led to the end of T-44 (1944 model) production in 1948 and the use of the tanks only in various units for training in the following years. Some of the vehicles ended up in the Soviet Army film studio, where they were still used well into the eighties.

Medium Tank T-44
Sredniy tank T-44 (obr. 1944 g.)

Made in: USSR (tank factories in Kharkov, Nishni Tagil, Omsk)
Used in: USSR
Developed: 1944
Manufactured: 1944-1948
Crew: 4 men.
Fighting weight: 31.5 tons
Overall length: 7650 mm
Hull length: 6090 mm
Width: 3140 mm
Height: 2410 mm
Armament: 1 85 mm ZIS-S-53 L/54.5 tank gun (1944 model) (88 rounds)
2 7.62 mm DTM machine guns (2750 rounds)
Powerplant: 12-cylinder W-2-44 Diesel engine (520 HP / 382 kW)
Top speed: 55 kph
Range: 235 km

The SPK-5 field repair crane was used by armored units of Russia's allies in the sixties.

Russian Tanks and Armored Vehicles 1946-to the Present

Although the development of the T-44 medium tank (1944 model) began in the penultimate year of the war, the tank only began to be introduced into a few tank regiments at the end of 1946.

During the course of 1944, the introduction of the IS-2 (1943 model) heavy tank and the modernized IS-2 (1944 model) into the units of the Red Army began. The latter type differed particularly in the thickness of the armor on the front of the hull and the installation of a 12.7 mm DSchK (1938 model) anti-aircraft machine gun. In the last war year the IS-3 (1945 model) heavy tank also appeared. While the Soviet Union's allies used the 1943 model IS-2 and 1944 model IS-2m practically unchanged until the end of the sixties, the Soviet Army's vehicles were equipped as of 1958 with a motor whose power was increased to 700 HP, a five-speed gearbox and a new radio set, and these tanks were used with the designation of Heavy Tank IS-2M (1958 model). As of 1960, the IS-3 was also fitted with the more powerful W-54/K-IS Diesel engine, and was designated Heavy Tank IS-3M (1960 model). Tests of the prototypes of the IS-4 heavy tank (1944 model, "Object 701"), IS-6 (1944 model) and IS-7 (1945 model, "Object 260") developed in 1944 and 1945, with various forms of armor and running gear continued until 1957-58. In addition, the power of the motors had been increased again and again. The IS-7 heavy tank (1948 model) had a 12-cylinder M-50 F3 Diesel engine producing 1250 HP (*see caption page 22*). The development of heavy tanks ended with the T-10 (1950 model). It was built from 1950 to 1957. As early as 1956 the production of the T-10M heavy tank (1956 model) began in Chelyabinsk. This version differed chiefly in the installation of a 14.5 mm KPWT heavy tank machine gun, in place of the 12.7 mm weapon, coaxially to the tank gun, and by the use of the 14.5 mm KPWT anti-aircraft machine gun in the loader's hatch.

A few mustered-out IS-2 hulls were still in use as heavy armored towing tractors years after the war.

Heavy Tank IS-4 ("Object 701-6")
Tyashelyi tank IS-5 (obr. 1947 g.)

Made in: USSR (tank factory in Chelyabinsk)
Used in: USSR
Developed: 1947
Manufactured: 1947-48
Crew: 4 men
Fighting weight: 60 tons
Overall length: 9790 mm
Hull length: 6600 mm
Width: 3260 mm
Height: 2480 mm
Armament: 1 122 mm D-25 tank gun (30 rounds)
           1 12.7 mm DSchK-M heavy machine gun (500 rounds)
           1 12.7 mm DSchK-M AA machine gun (500 rounds)
Powerplant: 12-cylinder W-12 Diesel enine (750 HP/552 kW)
Top speed: 43 kph
Range: 320 km

Heavy Tank IS-7
Tyashelyi tank IS-7 (obr. 1948 g.)

Made in: USSR (tank factory in Leningrad)
Used in: USSR
Developed: 1945-1947
Manufactured: 1947-48
Crew: 5 men
Fighting weight: 68.0 tons
Overall length: 11,480 mm
Hull length: 7380 mm
Width: 3400 mm
Height: 2480 mm
Armament: 1 130 mm S-70 tank gun (25 rounds)
           1 14.5 mm KPWT heavy machine gun (500 rounds)
           6 7.62 mm SGMT machine guns (6000 rounds)
           1 14.5 mm KPWT or 12.7 mm DSchK-M AA machine gun (500 rounds)
Powerplant: 12-cylinder M-50T (1050 HP/773 kW) or M-50F3 (1250 HP/920 kW) Diesel engine
Top speed: 55-59 kph
Range: 300 km

A Soviet armored unit being trained to attack with T-44M and T-54 medium tanks.

Chapter 2: The Russian Armored Vehicles of World War II

The IS-2 heavy tanks (1943 model) also spent many years in the armored troop units.

A great number of IS-2 heavy tanks (1943 model) were finally, after being mustered out, used as targets at training camps.

Heavy Tank T-10M ("Object 272 and 734")
Tyashelyi tank T-10M (obr. 1957 g.)

Made in: USSR (tank factories in Leningrad ("Object 272") and Chelyabinsk ("Object 734"))
Used in: USSR
Developed: 1950 (T-10) and 1956 (T-10M)
Manufactured: 1950-1962
Crew: 4 men
Fighting weight: 51.5 tons
Overall length: 10560 mm
Hull length: 7250 mm
Width: 3380 mm
Height: 2585 mm
Armament: 1 122 mm M-62-T2 tank gun (30 rounds)
           1 14.5 mm KPWT heavy machine gun (372 rounds)
           1 14.5 mm KPWT AA machine gun (372 rounds)
Powerplant: 12-cylinder W-12-5 Diesel engine (750 HP/552 kW)
Top speed: 50 kph
Range: 350 km

The (modernized) IS-2 heavy tank (1944 model) was fitted with reinforced front armor and a 12.7 mm DSchK AA machine gun (1938 model) and turned over to the armored troops.

The wreck of an IS-2(m) heavy tank (1944 model), seen at a troop training camp in the eighties.

From 1958 on, in the course of major overhauling, some IS-2 tanks were equipped with 700 HP Diesel engines, five-speed gearboxes and new radio sets (*see next page*) and turned over to the armored troops as IS-2M heavy tanks (1958 model).

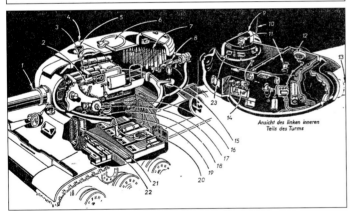

In the last war year, the IS-3 heavy tank (1944 model) appeared.

The SU-76m light self-propelled mount (1943 model) still used to the end of the fifties was already only conditionally useful in the last two years of World War II because of the insufficient penetrating power of its 76.2 mm regimental cannon. In addition, the vehicles often broke down from overheating of their two GAZ-203 gasoline engines. Since there were scarcely any spare parts available for the self-propelled mounts used by the Eastern Bloc states, they were, as a rule, used only as mobile points of fire. But some of these vehicles still remained in the reserves of these armies for a long time. Only in 1969 were the last 67 light SU-76m self-propelled mounts (1943 model) of the East German army taken out of the camps and, with only one exception, scrapped.

In the mixed tank and self-propelled gun regiments of the rifle divisions, the Red Army and their allies still used the SU-85 (1943 model) and SU-100 (1944 model) medium self-propelled mounts until 1964-65. While production of the SU-85 (1943 model) had been halted already in 1944, production of the SU-100 (1944 model)

The IS-3 heavy tank (1944 model) with DSchK AA machine gun (1938 model).

The IS-3M heavy tank (1960 model) was powered by the upgraded 700 HP W-54/K-1S Diesel engine.

Chapter 2: The Russian Armored Vehicles of World War II

The IS-4 heavy tank (1947 model) was developed as "Object 701-6" and produced from 1947 to 1948.

A tank driving school using IS-3M heavy tanks (1960 model).

Rear view of an IS-4 heavy tank (1947 model).

in the Soviet Union continued until 1946. In the same year, Czechoslovakia began production, continuing to produce the self-propelled gun mounts into 1947. After they were mustered out, some of the self-propelled mounts had their guns removed, were equipped with towing booms, and were used for many years as SU-85-T and SU-100-T armored towing tractors. A few armored tractors were still used by the Warsaw Pact nations until the beginning of the nineties. The ISU-122 (1943 model), ISU-122s (1943 model)—the latter reached the troops in 1944 and was fitted with a muzzle brake—and ISU-152 (1943 model) were still used some years after the war

Flamethrowing and chemical training at a Soviet engineer training camp.

IS-7 Heavy Tank

At the beginning of 1945, much emphasis was already being given to the further development of heavy tanks. Leading the development of the vehicle designated "Object 260" was S. J. Kotin, the chief designer at Factory No. 100 in Chelyabinsk. In the same year, the wooden model of the "Object 260" heavy tank (1945 model) was introduced. When Kotin and some of his colleagues were transferred in the spring of 1946 to the Leningrad tank works, which were to be modernized, they continued their developmental work there. After various changes, four hulls and four turrets, along with numerous individual parts, were made in Leningrad. The primary weapon used was the 130 mm B-13 cannon developed for the navy before World War II. Since the intended high-powered Diesel engine was not yet fully developed, the ACH-300 aircraft engine was revised as the TD-300 tank engine. For the first time, tracks with rubber bearings were used. In September 1946 the first prototype was finally driveable. Designated IS-7 Heavy Tank (1946 model), it was tested immediately. The second prototype was ready on December 25, 1946, and the remaining hulls and turrets were taken to the artillery test center in Kubinka for firing tests. As a result of the tests, the IS-7 ("Object 260") was reworked again. The lower hull sides were reinforced to 100 mm and the turret was redesigned. To hold the 1050 HP M-50T Diesel engine, which was based on the M-50 naval Diesel, the hull was widened. For reasons of space, water-cooling by means of the usual bucket coolers had been ruled out and the pressure of the exhaust gas utilized instead. In the summer of 1948, four IS-7 heavy tanks (1948 model) were ready for testing. Although one of the vehicles caught fire and burned out during firing tests, another showed poor workmanship in the engine room during firing tests, and numerous improvements were required, the Soviet defense industry considered the creation of fifty pre-series vehicles. By 1949, though, troop testing had revealed the advantages of the T-54 medium tank, so the decision had to be made to modernize the heavy tank but not to change it thoroughly. As a result of political disagreements in Leningrad in 1949-50, Kotin finally returned to Chelyabinsk and reworked the IS-3 heavy tank.

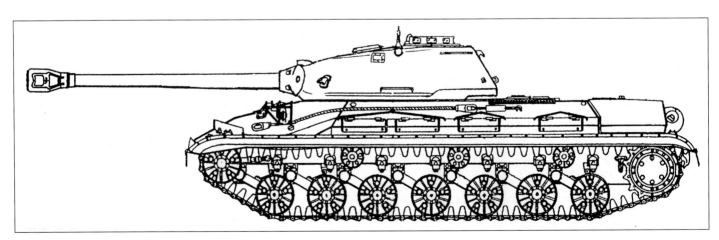

Improving the heavy tanks continued at Chelyabinsk under the designation "Object 730." Kotin was again in charge. In 1950 the first prototype was completed. The pre-series vehicles underwent troop testing as IS-8 heavy tanks (1950 model). In the process of revising the Stalin personality cult, the IS-8 (IS = Josef Stalin) were now designated T-10.

Chapter 2: The Russian Armored Vehicles of World War II

In the first half of the fifties, the 122 mm D-25TA tank guns were fitted with a stabilizing system and used in heavy tanks now known as T-10A. At about the same time, a smoke ejector for the gun was also developed.

In the course of modernization, the T-10 heavy tanks (1950 model) were also equipped with an infra-red night firing and night vision system.

With the introduction of the upgraded 122 mm M-62-T2 L/54.6 tank gun (recognizable by the new muzzle brake), the T-10 heavy tank was given the suffix "M."

The T-10M heavy tank (1956 model), seen in August 1961 during the sealing off of Berlin.

Right: Production of the T-10M heavy tank (1956 model) ran from 1957 to 1962 at the Chelyabinsk and Leningrad tank works. About 8,000 tanks were produced.

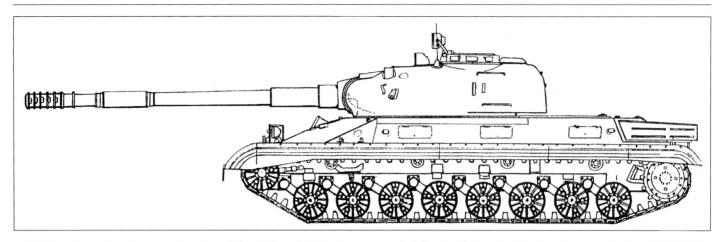

In 1957 another attempt was made to install the 130 mm B-13 tank gun reworked for the IS-7 and a Diesel engine upgraded to 1000 HP. The tests took place with the "Object 277" (1957 model) prototype.

in the Soviet army, and by several allied states until at least the mid-eighties. The production of the ISU-152 (1943 model) continued until 1947. The ISU-152 vehicles used in the Soviet Army were later fitted with 700 HP W-54/K-IS Diesel engines and kept in use as ISU-152K (1955 model).

A good many of the heavy self-propelled mounts had their howitzers removed after they were taken out of service. They were then fitted with towing apparatus and used as ISU-T heavy towing tractors, and as of 1962 designated BTT-1 or BTT-1T (Bronetankowiy tyagashelyi); these too saw service until the nineties.

During World War II the Russian armament industry produced only one version of an armored scout car. These vehicles, designated Armored Car BA-64 (1942 model), were still used by numerous allies to the end of the fifties; some of them were also used as towing tractors for light guns and grenade launchers. A few vehicles had their turrets removed and a rear door installed and were then used as armored troop carriers. At the beginning of the eighties, small numbers of these armored cars were still being used in Albania and North Korea. Besides, various international references mention that a considerable number of ISU-152s are supposed to have been equipped with 520 HP diesel engines W-54 between 1959 and 1964, and that they were used in the Soviet Army.

Although it proved to be an insufficient antitank weapon in World War II, the SU-76m light self-propelled gun (1943 model) was still used after the war by the Soviet Union and other forces.

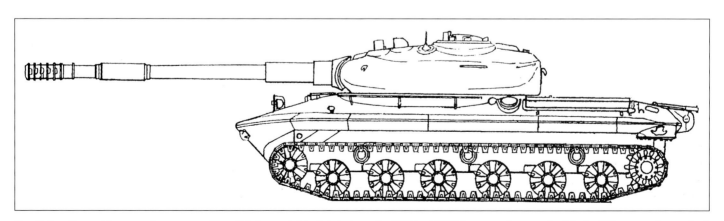

To attain a favorable ground pressure with a fighting weight of sixty tons, the "Object 279" heavy tank, armed with the 130 mm tank gun, had four track drives.

Chapter 2: The Russian Armored Vehicles of World War II

Cleaning the 76.2 mm ZIS-3 regimental cannon in its self-propelled mount.

The left half of the fighting compartment of the SU-76m (1942 model) light self-propelled gun.

Side view of one of the few surviving SU-76m (1942 model) guns.

SU-85 (1943 model) and 57 mm ZIS-2 antitank gun (1943 model) used in a drill in the fifties.

Like the T-34 hulls, numerous chassis of the SU-85 (1943 model) were used as SU-85-T medium armored tractors.

Left: To motorize the rifle divisions in the shortest time, a mixed armored self-propelled gun regiment was subordinated to them. In the first years, the SU-85 medium self-propelled gun (1943 model) was among their equipment.

Rear view of a T-34-T medium armored tractor, rebuilt and used by the East German People's Army.

The mixed regiments were also supplied with SU-100 medium self-propelled guns (1944 model) built in World War II.

The SU-100 medium self-propelled gun (1944 model) with a mixed armored/self-propelled gun regiment stationed in East Germany.

Production of the SU-100 medium self-propelled gun (1944 model) went on in the USSR until 1946. In that year the CSSR took over production of these vehicles, which was halted a year later.

Above Left: A mixed regiment, equipped with SU-100 (1944 model) guns, on the march. *Above Right*: The SU-100 (1944 model) was armed with the 100 mm D-10S L/56 tank gun, which was ultimately modified slightly and used as the D-10T tank gun in the T-54.

Chapter 2: The Russian Armored Vehicles of World War II

Since 1947, the five-speed gearbox of the T-34-85 (1947 model) has also been used for medium self-propelled guns. To differentiate these vehicles, they were designated SU-100M (1947 model).

A 1968 photo taken at a tank driving school. Among various modern armored vehicles, there is still one SU-100M (1947 model) in use.

This SU-100M medium self-propelled gun (1947 model) was used by the Hungarian Army after World War II.

On display at the museum in Berlin-Karlshorst is this ISU-152K heavy self-propelled gun (1955 model).

Production of the ISU-152 heavy self-propelled gun (1943 model) continued until 1947. Here a regiment equipped with the ISU-152 (1943 model) is seen on parade in May 1953.

The self-propelled mounts based on the IS chassis were also rebuilt as ISU-T heavy armored tractors.

As of 1955, the first ISU-152K heavy self-propelled guns were supposed to be armed with the upgraded 700 HP W-54/K-1S Diesel engine. The vehicles were recognizable by the reserve oil tank on the right front track apron.

Right: The ISU-T heavy armored tractor of an armored unit of the Soviet Army in the sixties.

During the Hungarian uprising in 1956, the Soviet Army used the ISU-152K heavy self-propelled gun (1955 model) to put down the demonstrations.

For some years after the war, the BA-64 armored car (1942 model) was used by the Soviet Army's reconnaissance units. The Albanian and North Korean forces even used them until the early eighties.

Chapter 2: The Russian Armored Vehicles of World War II

The BTT-1 heavy armored tractor (1962 model).

The BTT-1T heavy armored tractor (1962 model).

The only light tank of the postwar era

In addition to the aforementioned K-90 prototype, the Russian arms industry has built, from the end of World War II to the present day, just one model of a light armored vehicle. This had been developed from 1948 to 1951 by the Volgograd tank factory, on the basis of an Antarctic tracked tractor, as an armored reconnaissance car, and was particularly unique for its time in its combination of relatively large-caliber armament and good amphibian capability. Its high degree of maneuverability in the water was attained by having a water-jet powerplant installed in the rear of the hull. Water was sucked in through two openings on the underside of the hull and forced backward by a pair of shovel-wheel turbines. The greater the pressure and thereby the required quantity of water, the greater was the pressure, too, and thus the speed.

With the possibility of being able to steer the two jets differently, the direction of floating in the water could be influenced. It was introduced under the designation "PT-76 light amphibian tank (1951 model)." In standing waters the tank reached a top speed of up to 10 kph. In the first production run of the amphibian tank the 76.2 mm D-56T (1954 model) tank gun was still fitted with a multichambered muzzle brake. The tank delivered to the troops as of the mid-fifties as the light amphibian tank PT-76 (1954 model) contained the improved D-56TM tank gun with a two-chambered muzzle brake.

In the modernized tank, now PT-76 light amphibian tank (1958 model), there was, along with numerous very small technical modifications, the "Sarya" stabilizer installed as an important new development. Thereby, with the stabilizing of the tank gun in horizontal and vertical planes, accurate firing became possible, even when in motion at slow speeds. At the same time, a traverse indicator with greater diameter had been installed, in order to show the gunner the direction of travel more easily while firing. A new line plate was installed in the TSchK-66K turret aiming scope. This now had a scale for hollow-charge shells and another by which targets 2.70 meters above ground could be estimated in terms of distance. Through the introduction of an infra-red night-vision apparatus, the

A BA-64 armored car (1942 model) used for reconnaissance.

The BA-64 was armed only with a 7.62 mm DT tank machine gun.

The PT-76 amphibian tank (1951 model), seen during Soviet army attack training.

The PT-76 light amphibian tank (model 1951) was armed with the 76.2 mm D-56T tank gun and had a multichambered muzzle brake.

A reconnaissance unit, equipped with PT-76 (1951 model), seen in a 1951 parade in Moscow.

The PT-76 light amphibian tank (1954 model), equipped with the 7.62 mm D-56TM tank gun and two-chambered muzzle brake, seen during a sea-landing drill.

tank was also suitable for night action. Thus, the commander, who was also the aiming gunner, had the TPKU-2 observation device and the TKN-1 infra-red observation device (TKN-1S as of 1964) at his disposal. For the driver, the TWN-2B infra-red observation device was installed. The TNP-370 driver's periscope was replaced by the PER-17A. The driver's other two angle mirrors were, as before, of the TNP type.[8]

The amphibian tanks built until 1967 were equipped with the R-113/26 tank radio set and the R-120 on-board speaker set. The vehicles built after that had the R-123 tank radio and the R-124 speaker.

Amphibian Tank PT-76

Plavayushiy tank PT-76 (obr. 1951 g.), PT-76 (obr. 1954 g.), PT-76B (obr. 1958 g.)

Made in: USSR (tank factory in Volgograd)
Used in: USSR, Afghanistan, Algeria, Angola, Benin, Bulgaria, China, Congo, CSSR, Cuba, East Germany, Egypt, Guinea, Guinea-Bissau, India, Indonesia, Iraq, Israel, Laos, Madagascar, Nicaragua, North Korea, Pakistan, Poland, Syria, Uganda, Vietnam, Yugoslavia, Zambia.
Developed: 1951
Manufactured: 1952-1974/75
Crew: 3 men
Fighting weight: 14.0 tons
Overall length: 7625 mm
Hull length: 6910 mm
Width: 3140 mm
Height: 2195 mm
Armament: 1 76.2 mm D-56T or D-56TM tank gun (40 rounds),
1 7.62 mm SGMT or PKT machine gun (2000 rounds)
Powerplant: 6-cylinder W-6 Diesel engine (240 HP/177 kW)
Top speed: 44 kph (road), 10.2 kph (water)
Range: 250 km

---

8. USSR Defense Ministry, Handbook for PT-76 Tank, Moscow 1963.

Chapter 2: The Russian Armored Vehicles of World War II

After the reconnaissance units of the armored and motorized rifle divisions had been supplied with the BMP, the PT-76 was only used by the naval infantry units of the Soviet forces.

In 1960 the Volgograd tank factory suggested a new shape for the hull of the amphibian tank. The reason why the development of the PT-76M (1960 model) amphibian tank was stopped may have been that despite the changed hull, the speed could be increased only insignificantly, while the total weight had been increased.

In hopes of also increasing the penetrating power of the tank guns of the amphibian tanks, the "Object 906" light amphibian tank, armed with an 85 mm tank gun, appeared in 1963.

When the introduction of the BMP-1 armored troop carrier in the early seventies marked the end of PT-76 production, the Russian tank designers in Volgograd proposed two successor models. They were introduced in 1975 as the "Object 934" light amphibian tank and the "Object 685" light airborne tank, which could be used amphibiously. Although the installation of a 100 mm tank gun produced considerable firepower for amphibious armored vehicles, this development was not pursued further.

When practice cartridges were fired, the muzzle brakes were usually removed, since they showed no effect and only got dirty from the powder smoke. This PT-76B (1958 model) was used by the Polish army.

Winter training with a PT-76B light amphibian tank (1958 model) in February 1965.

In the water, the PT-76 was driven by two jet motors. They sucked water in under the hull and drove it out at the rear. Photo taken July 19, 1968.

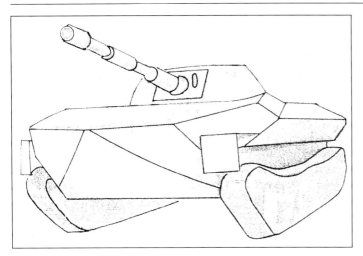

Only a prototype of the PT-76M (1960 model) amphibious tank, proposed with a new armored hull, was built.

The T-54/55 Medium Tank

At the end of 1945 the Russian tank designers succeeded in developing a new, wider turret, which made it possible to install a 100 mm tank gun and a parallel-barreled machine gun. In addition, another machine gun was installed on each side of the hull on the forward track apron. Just one year later, production of the first series of the T-54 Medium Tank (1946 model) began at the Kharkov Tractor Works. The T-54 Medium Tank (1949 model) produced as of 1949 already had the typical hemispherical shape. Its rear sides, though, were still drawn upward. In addition, only the coaxial SGMT turret machine gun was installed along with the tank gun.

In order to reduce the places where shells could lodge, especially on the turret, a turret with further modifications was used on the T-54 (1951 model) Medium Tank. In the last year of the war, hemispherical turrets were already being used on heavy tanks.

The T-54 Medium Tank (1951 model) that first appeared during the uprising in Hungary attracted great international attention. The following evaluation, indicative of the author's impression, was stated in a contemporary report published in West Germany: ". . . the T-54 is a fruit of long years of development, which was based on the true and tested design elements of the T-34 series. The Soviets have succeeded, through outstanding space utilization, weight reduction, and a turret shape that could be called a work of genius, in mounting a 100 mm tank gun on a 38-ton vehicle. Thereby the tank is equal or superior to all other types up to 50 tons that are being used by troops at this time.

The high achievement of the 100 mm tank gun can be augmented by refined fire control systems, quite apart from the facts that these are generally more prone to disturbance and require considerably more training. The operative mobility of the T-54 is unsurpassed. Its high range makes it independent of any fuel supplying in battle for a distance of some 350 kilometers. Units not in combat could drive from the zone border to the Rhine without refueling! The simple design of the running gear, powerplant and armament made it rugged and independent of servicing. It also made mass production easier. The required work hours are estimated to include only 25% of the time usually needed in the West. The tactical mobility is determined by the power-to-weight ratio of 14.5 hp/ton. The extremely low silhouette also benefits mobility on the battlefield. In addition, the tank has the general advantages of the Diesel engine: low noise production, low heat development in the exhaust system to make infrared detection difficult. These advantages make a comparatively light armor possible. It is strengthened as much as possible by the ideal shape, especially that of the turret. Thus the advantages of the T-54 as compared with the heaviest Western types of the 45- to 50-ton class (M48 and Centurion) can be regarded above all in the following order: superior armament, superior operative and tactical mobility, little chance of breakdown, thus greater logistic independence, simplified production and simpler training requirements. . . . At this time the West has nothing equal with which to oppose the T-54."[9]

To this day, the T-54 medium tank is found among the weapons of numerous fighting forces. In 1991 the Russian Defense Ministry stated that there were still 1,379 of these tanks among the weapons of the Soviet Army. But forty years of service resulted in the fact that numerous technical innovations had found their way into the old tanks, step by step. Under pressure to give every little change a new designation, the experts of the so-called specialist realm came up with the strangest type designations. The Soviet tank designers may have observed this with smiles on their faces, especially when, what with the conflict of rearmament and new developments, some real blunders were made. It stands to reason that, with an average yearly production of more than 2,000 tanks, not every innovation can be added to every available tank. In fact, in the Soviet training and maintenance directives, all T-54, T-55 and T-62 types were combined simply as medium tanks.

Among the innovations added step by step and to varying extents were:

- Elimination of the SGMT 7.62 mm tank machine gun formerly included in the hull,
- Fitting with equipment for underwater driving,
- Inclusion of stabilization for the D-10TG L/56 tank gun and turret machine gun,
- Inclusion of a smoke ejector at the mouth of the barrel,
- Extension of the scale in the turret targeting scope for low-caliber shells,
- Improvements to the WUK-27A turret-turning apparatus,
- A new collecting-ring transmitter,
- Higher-performance G-74 generator with RRT-31A voltage regulator,
- Improved air filter with two-stage operation, and all modernizations that applied to the successor T-55 type.

All the modernized T-54 tanks simply had the letter "M" added, with the exception of the vehicles that were built during the conversion to T-55 production in 1954-55. These tanks corresponded greatly to the T-54, but had the STP-1 "Horizon" stabilizing system and a smoke ejector. The command tanks equipped with two radio stations were also differentiated. They bore the designation of Medium Tank T-54K, T-54AK or T-54MK. The versions built after 1957, with lateral and elevating stabilization for the D-10T 100

---

9. Senger & Etterlin, F.M. von, *Die roten Panzer*, Munich 1963, pp. 251-252.

Chapter 2: The Russian Armored Vehicles of World War II

The T-54 medium tank (1946 model) was built from 1947 to 1949. The BTU-55 dozer hitch (1963 model) was attached to this T-54 tank.

The T-54 (1949 model) successor version already had the typical "turtle shape" in the front half of the turret. The rear of the turret, though, was extended upward.

mm tank gun and without a smoke ejector had to have a reinforcing ring added at the gun opening to make up for the added weight in the front end. This version was designated Medium Tank T-54B (1957 model). Gradually, though, these tanks became T-54M medium tanks. A number of "B" versions were still in action in the nineties. In 1991 it was stated that they were still in use.

Medium Tank T-54, T-54A, T-54M, T-54B
Sredniy tank T-54 (obr. 1951 g.), T-54A (obr. 1954 g.), T-54M, T-54B (obr. 1957 g.)

Made in: USSR (tank factories in Kharkov, Nishni Tagil, Omsk), Czechoclovakia (Dubnica, Nowi Jicin), Poland (Katowitze, Labedy), China
Used in: USSR, Afghanistan, Algeria, Angola, Bangladesh, Bulgaria, China, Congo, Cuba, Czechoslovakia, East Germany, Egypt, Ethiopia, Finland, Guinea-Bissau, Hungary, India, Iraq, Israel, Laos, Libya, Madagascar, Mali, Morocco, Mozambique, Nicaragua, Nigeria, North Yemen, Pakistan, Peru, Poland, Romania, Somalia, South Yemen, Sudan, Syria, Tanzania, United Arab Emirates, Vietnam, Yugoslavia, Zambia, Zimbabwe.
Developed: 1950
Manufactured: 1951-1955 (USSR), 1958-1963/64 (CSSR, Poland)
Crew: 4 men
Fighting weight: 36.0 tons
Overall length: 9000 mm
Hull length: 6450 mm
Width: 3270 mm
Height: 2400 mm
Armament: 1 100 mm D-10TG L/56 tank gun (34 rounds)
           1 7.62 mm SGMT machine gun (3500 rounds)
           1 12.7 mm DSchK-M AA machine gun (500 rounds)
Powerplant: 12-cylinder W-2-54 Diesel engine (520 HP/382 kW)
Top speed: 50 kph
Range: 450 km

The T-54 medium tank (1949 model) was produced until 1951.

The T-54A medium tank (1951 model) was now equipped with a hemispherical turret.

The T-54 medium tank (1946 model) with dummy gun shield, used in training close-combat soldiers.

Not until 1954 was the basic revision of the medium tanks completed. The successor to the T-54 was produced in series in the Soviet Union as Medium Tank T-55 (1954 model) from 1955 to 1960. Externally it could be recognized by the lights of the infra-red night vision apparatus and the lack of an opening on the turret roof for ventilation of the fighting compartment. The latter was connected with an improved weapons protection apparatus and the related stronger filter ventilation system for used in contaminated combat areas. In addition, the following innovations were introduced:
- Full stabilization of the improved 100 mm D-10T2S L/56 tank gun through the installation of the STP-2 "Zyklon" electro-hydrostatic stabilizer,
- Addition of a rotating stage under the tank gun,
- Introduction of a more powerful smoke-screen system,
- Use of the 580 HP W-2-55 Diesel engine,
- Installation of a compressed-air-supported coupling and power steering (previously the steering had been supported by a servo spring),
- Use of a lower lateral power train,
- Enlargement of the fuel tank, and a
- Combat supply increased by nine shells.[10]

At the end of the fifties, the military leadership of the USSR had come to the conclusion that aircraft had attained such high speeds that they could no longer be fought with a tank's anti-aircraft machine gun. Thus, a number of T-55 medium tanks were delivered

The T-54B medium tank (1957 model) equipped with vertical and horizontal stabilizers had to have a reinforcing ring attached to the muzzle of the tank gun as a counterweight. When reequipped with an infra-red night-vision device, this tank finally became the T-54M.

with a simple loader's visor. With the advent of the helicopter as a tank destroyer, the equipping of the loader's aperture with a turning-ring mount for the 12.7 mm DSchK-M (1938-46 model) anti-aircraft machine gun was resumed as of 1964-65. The tanks built up to then without the AA machine gun were reequipped step by step.

Barely ten years later, the tank was modernized into the T-55A, with the following improvements:
- Replacement of the SGMT tank machine gun by the PKT,
- Attachment of anti-radiation protective plates in the fighting and driver's compartments, as well as in the commander's, loading gunner's and driver's visors.
- Installation of hydraulically supported power steering,
- Use of the R-123 tank radio set and R-124 speaker system,
- A more powerful generator with 6.5 kW power and transistor-directed R-10T regulator,
- A new track-tension device,
- A main clutch increased by two to 19 clutch plates,
- A steering clutch improved to 17 clutch plates, and

June 21, 1968: a T-54A medium tank in use at a tank driving school.

10. USSR Defense Ministry, Textbook for NCOs in Armored Units, Moscow 1975.

Chapter 2: The Russian Armored Vehicles of World War II

The T-54A medium tank (1951 model) with BTU-55 dozer hitch is seen crossing the bridge of a MTU bridgelaying tank.

- Installation of a stronger compressor for the compressed-air system.[11]

As of the mid-seventies, new rubber lubricated track links with appropriately modified drive rings were used on medium tanks. The T-55AK command tank was also equipped with a radiation warning system.

Between 1963 and 1980, the Czech and Polish armament industries participated in the production of medium tanks, turning out more than 600 per year. The tanks built in the CSSR were designated T-55AM, those in Poland were called T-55A(P). The Polish tanks had a modified steering system, a different attachment of the air-filter hatch on the rear of the tank, and a generator with only 5 killowats and a suitable regulator.

As of 1976, the first medium tanks were equipped with a laser range finder. According to a list of available tanks, made available by the Soviet delegation at the Vienna Conference of 1991, these were called T-55AD. Officially, though, they were known to the Russian fighting forces only as Medium Tank T-55A (1976 model).

This T-54A (recognizable by the covering on the fighting-compartment ventilator) was also modernized at last into the T-54M. These were called T-55 in the armies of other Warsaw Pact states.

A significant increase in combat value was gained by the medium tanks from 1980 on, when they were modernized by several installations and additions. There was also the Volna computer-directed fire-control system, with laser range, speed, temperature (outside, inside, and loading area) and air-pressure indication, a new 2B-32K aiming telescope, the improved STP-2A stabilizing system, the Kladivo ("Hammer" in German) laser warning system, a 902B "Tutscha" foglaying system consisting of eight tubes, additional armor plate, and the R-173 tank radio set.[12] In the Soviet Army the vehicles with enhanced combat value resulting from the 1983 modernization were designated Medium Tank T-55AM (1980 model) and Command Tank T-55AMK (1980 model).

To remove mine barrages, the T-54 medium tank could be fitted with the KMT-4 mine removal device (or "roller section").

---

11. Study materials of the Military Academy, Modernization of the T55, Dresden 1974.
12. Ministry of National Defense, Fire Control and Laser Warning Department, Berlin 1988.

To cross water barriers, an air pipe was used in place of the loader's optics. This T-54A medium tank (1951 model) of the Polish Army has the 100 mm D-10TG tank gun installed.

Installing a turret on the T-54 in the Katowitze tank works in Poland.

Parallel to the introduction—begun in 1980—of extensive modernizing, the development of guided weapon complexes for firing rockets began at the Tulamashsavod machine works in Tula. Three years later, the 3UBK-10-1 antitank missile could be fired from the 100 mm tank gun of the T-55. In addition, instead of the TPN-1-22-11 night aiming scope, the 1K13 aiming scope complex of the 9K116 "Bastion" guided weapon system (NATO code: AT-10/STABBER) could be utilized. With 9M117 rockets, armor plate of 550 to 650 mm could be penetrated, even at a distance of 4,000 meters.[13]

As of 1986, the CSSR, the DDR and Poland also took part in manufacturing parts of the new equipment and remodeling the available T-55 tanks. In the Warsaw Pact states, these tanks were now known as T-55 AM2 or, if they had the guided weapon complex, as T-55 AM2B.

The process of equipping the available tanks with reactive added armor began in 1989. They were designated Medium Tank T-55MW and T-55AMW and are still in use. At least those tanks exported to other countries will still remain in military use for several years.

All versions of the T-55 were also delivered as command tanks. In the Soviet army they were given the suffix letter K. There were three different versions made for different levels of command. While the vehicles of company chiefs received only an additional radio set with a range of 40 kilometers, those made for battalion commanders had a telescopic mast that increased the range to 70 kilometers. The tanks for the commanders of armored regiments were fitted with the R-130 radio as a second station. To provide electricity, a generating aggregate was installed in the fighting compartment. While the first two versions had to have their ammunition supply reduced from 43 to 38 shells, the regiment commanders' tanks had only 26 rounds.

A small number of T-55 Medium Tanks were fitted with the ATO-200 flamethrower in place of the turret machine gun. These were in troop use as Flamethrowing Tank OT-55. A typographical error in the Russian military journal "Woenniye snaniya 8/84" was the reason why they are still known internationally by the designation of TO-55.

According to international estimates, more than 50,000 T-54/55 tanks have been built since 1951. Production continued longest at the tank factory in Omsk. Only in 1981 did the last T-55 medium

Cutaway model of the 12-cylinder, 520 HP W-54 Diesel engine. In the background is a T-55AM2.

---

13. Ministry of National Defense, Guided Weapon Complex 9K116, Berlin 1989.

Chapter 2: The Russian Armored Vehicles of World War II

The successor to the T-54 was the T-55 medium tank (1954 model); among numerous improvements, it had an infra-red night-vision apparatus. The first vehicles of this series still lacked AA machine guns when they were sent to the troops.

Winter training for the crew of a T-55 (1954 model).

The air pipe of the underwater equipment was attached to the rear of the T-55 medium tank (1954 model). The angle iron attached behind it was to protect the pipe from damage.

Shooting practice with the fully stabilized 100 mm D-10T2S tank gun of the T-55 (1954 model).

For underwater driving, the L-2g infra-red firing spotlight was removed.

As this picture was taken, the T-55 (1954 model) with BTU-55 dozer (1963 model) was just breaking its left track.

The T-55 medium tank (1954 model) with KMT mine removal device.

The KMT-5 mine removal device, seen here in August 1969, replaced the earlier model.

As of 1964-65, the T-55 was fitted again with the 12.7 mm DSchK-M AA machine gun (1938-46 model). With the other modernizations introduced around that time, the tank was designated as T-55A.

For protection from hollow-charge ammunition fired from airplanes and helicopters, the T-55A was fitted with light additional armor ahead of the turret hatches and in the center of the turret roof.

The splinter guard of the TPN-1-22-11 night aiming scope. Before it could be used at night, the cover had to be removed from it.

Chapter 2: The Russian Armored Vehicles of World War II

The T-55A medium tank. Hard blows to the individual torsion bars often caused them to break.

As of 1963, the T-55 was also produced in the CSSR, where it was designated T-55AM. The vehicle shown here bears chassis number MHY A 40 HKP and was thus the fortieth tank built in May 1969.

tanks leave the assembly halls there. The CSSR and Poland had already halted the production of these tanks a year before. According to Soviet reports, there were still 3,140 of these tanks in use by the Russian forces in 1991.

Medium Tank T-55, T-55A
Sredniy tank T-55 (obr. 1954 g.), T-55A (obr. 1964 g.)

Made in: USSR (tank factories in Kharkov, Nishni Tagil, Omsk), CSSR (Dubnica, Novi Jicin), Poland (Katowitze, Labedy), China
Used in: USSR, Afghanistan, Albania, Algeria, Angola, Austria, Bangladesh, Belgium, Bulgaria, Cambodia, China, Congo, Cuba, Czechoslovakia, East Germany, Egypt, Ethiopia, Finland, Guinea-Bissau, Hungary, India, Iraq, Israel, Madagascar, Mali, Mongolia, Morocco, Mozambique, Nicaragua, Nigeria, North Yemen, Pakistan, Peru, Poland, Romania, Somalia, South Yemen, Sudan, Syria, Tanzania, Togo, USA, Vietnam, Yugoslavia, Zambia, Zimbabwe
Developed: 1954
Manufactured: 1955-1981 (USSR), 1963-1985/86 (CSSR, Poland)
Crew: 4 men
Fighting weight: 36.0 tons
Overall length: 9000 mm
Hull length: 6450 mm
Width: 3270 mm
Height: 2400 mm (with AA MG 2840 mm)
Armament: 1 100 mm D-10T2S L/56 tank gun (43 rounds)
  1 7.62 mm SGMT or PKT machine gun (2250 rounds)
  1 12.7 mm DSchK-M AA machine gun (500 rounds)
Powerplant: 12-cylinder W-2-55 or W-2-55A Diesel engine (580 HP/427 kW, later 620 HP, 456 kW, finally 680 HP/500 kW)
Top speed: 50 kph
Range: 500 km

The T-55 was also produced in Poland. To tell it apart from the original version, it was given the suffix "(P)." This T-55A(P) is forcing a passage through a water barrier, with a Star truck and 85 mm anti-tank gun in tow, during the "Weapon Brotherhood" maneuvers of the Warsaw Pact states, October 12-18, 1970.

A tank unit of the Czech Army, equipped with T-55 tanks, during the "Weapon Brotherhood 80" maneuvers.

A T-55A medium tank with BTU-55 dozer (1963 model).

The KMT-6 ("knife section") mine removal device consisted of two plows which were supposed to bring the mines out of the ground and push them aside. This East German T-55 with KMT-6 was photographed in October 1970.

A T-55A with BTU-55 in action in the winter of 1978-79.

Usually the "roller" and "knife" sections of the mine removal device were used simultaneously.

Chapter 2: The Russian Armored Vehicles of World War II

The T-55A medium tank with KMT-6 mine removal device.

As of 1976, the first T-55 tanks were fitted with a laser range finder above the 100 mm tank gun.

In the Soviet Army, the tanks equipped with laser range finders were called T-55A (1976 model). At the disarmament conference they were designated T-55AD.

After 1980 another far-reaching modernization of medium tanks was begun. This included, among other things, a calculator-guided fire-control system, an integrated laser range finder, a laser warning system, a foglaying device, a guided weapon complex and additional armor plate. In the Soviet Army these tanks served as T-55AM (1980 model). Since this designation had been used already in the Warsaw Pact states, they were called T-55AM2 there. After the introduction of the 9K116 "Bastion" guided weapon complex they became T-55AM2B.

The laser range finder above the 100 mm D-10T2S tank gun and the L-2g firing spotlight of a T-55AM (1980 model), here serving in the East German army as T-55AM2B.

The 902B "Tutsha" foglaying apparatus consisted of eight launchers.

The "Klavido" (hammer) laser warning system of the T-55AM (T-55AM2B).

Some T-55s were fitted with flamethrowers in place of the turret machine gun.

Chapter 2: The Russian Armored Vehicles of World War II

The OT-55 Flamethrowing tank (1956 model) was built from 1957 to 1962. The original Russian designation was Ognemetnyi tank OT-55 (obr. 1956 g.)

In the mid-fifties, a 100 mm smooth-bore tank gun is said to have been installed in the T-55. Attempts were made under the designation of "Object 140." This photo may show one of those tanks, for in this typical T-55 turret with added laser range finder, there is neither a smoke ejector nor a reinforcing ring.

The T-62 Medium Tank

When the medium tank with the 100 mm tank gun was introduced, it could already be seen that this caliber would not suffice in the long run. But since an increase in caliber inevitably resulted in a heavier gun, that added recoil power which affected the entire turret, countless firing tests were used to determine the upper limits of the caliber, which increased the mass of the barrel only to a slight extent. Although the Russian standard caliber of 122 mm originally might have been intended for heavy tanks, the realizable maximum appears to have been 115 mm. Through the use of a smooth barrel and wing-stabilized ammunition, the penetrating power of the hollow-charge shells could also be increased significantly. During World War II it was already learned that the penetrating power of a hollow-charge shell can equal six times the caliber if the shell hits the target without rotating. The quick development of Russian guided antitank missiles in the late fifties thus benefited tank guns as well.

This was recognized by Western tank designers only in the early seventies.

In the mid-fifties, testing of the new 115 mm D-68 tank gun began. For it, two T-54 tanks were fitted with this weapon. The Object 165 tanks used the conventional T-54 turret. For the prototype, designated Object 166, a new turret design was used. Ultimately this prototype resulted in the T-62 tank. The 115 mm D-68 tank gun was introduced into the armaments of the Soviet forces as U-5TS (2A20).

In 1961 the development and testing of the T-62 medium tank (1960 model), begun in 1957, was finally concluded; a pre-series of 25 vehicles had been produced from the end of 1961 to mid-1962. Large-series production was begun in July 1962. The command tanks were fitted with a second radio set and had the letter K added to the type designation. The first tanks to be built were similar to the T-55 medium tanks built at the end of the fifties, in that

In the mid-fifties, the first experiments with enlarged-caliber tank guns began. The tests were carried out with "Object 165." For this, the 115 mm D-68 tank gun was installed in a typical T-54.

Since enlarging the tank gun's caliber required a modified turret, a new turning turret was designed and mounted on the hull of a T-54. The resulting vehicle was called "Object 166."

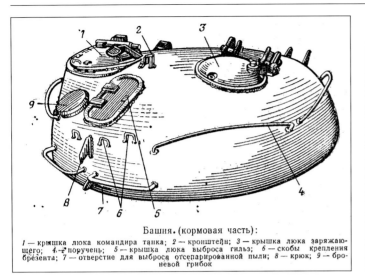

Diagram of the modified tank turret on "Object 166."

The T-62 medium tank (1960 model) with the 115 mm tank gun introduced as U-5TS (2A20).

they had no anti-aircraft machine gun. But in this tank type, too, as of 1965 the 12.7 mm DSchKM anti-aircraft machine gun (1938-46 model) was reintroduced. For this the loader's side of the turret was designed completely anew. This version was known officially as Medium Tank T-62 (1965 model). No rearmament of the turrets that were not equipped with this machine gun has been done to this day. The improvements that had already appeared in the T-55 Medium Tank were added to this tank gradually, though only in part. All of these tanks were uniformly designated Medium Tank T-62M or T-62MK. In 1976 the first tanks of this type were also equipped with a laser range finder mounted over the tank gun. They received the designation of Medium Tank T-62M (1976 model). By 1972 production added up to some 20,000 tanks. Twenty countries of the world have been equipped with these tanks to date.

As of 1980, the T-62M was surely supplied with the same modernizations as the T-55. These tanks became known internationally primarily from their use in Afghanistan. The T-62 tanks used there were, for the most part, equipped with the laser range finder above the tank gun and the additional armor on the front of the hull and turret as well as on the sides. In Western literature they are designated T-62 E. At the same that rockets were developed for the 100 mm tank gun, guided missiles were also made for the U-5TS (2A20) tank gun of the T-62. Here, too, the night aiming scope was replaced by the 1K13-1 targeting scope complex of the 9K116-1 "Sheksna" guided missile system. The ammunition was introduced as 3-UBK-10-2.

Although no information about the equipping of the T-62 with reactive armor has been found, it can be assumed that T-62MW versions have existed.

Externally, the armored hull of the medium tank introduced as T-62 (1960 model) scarcely differed from its forerunner. The different distance between the road wheels came about only because the enlarging of the servicing hatch under the engine. In the first series-production vehicles, the L-2g infra-red spotlight was still mounted above the tank gun.

The turret of the T-62 had a considerably larger diameter, which decreased the space on the two track aprons.

Chapter 2: The Russian Armored Vehicles of World War II

Refueling a tank company of T-62 (1960 model) tanks in the field.

The KMT-4 ("knife section") mine removal device could also be mounted on the T-62.

A number of the medium tanks were fitted, instead of with the 7.62 mm PKT machine gun in the turret, with the flamethrower that was also used in the OT-55. These vehicles were used by the Soviet forces as OT-62 Flamethrowing Tanks.

In 1991 the Russian government still reported the presence of 2,144 T-62 Medium Tanks in their fighting forces.

Medium Tank T-62, T-62M
Sredniy tank T-62 (obr. 1960 g.), T-62 (obr. 1965 g.), T-62M (obr. 1976 g.)
Made in: USSR (tank factories in Kharkov, Omsk, Nishni Tagil)
Used in: USSR, Afghanistan, Algeria, Angola, Bulgaria, Chad, Congo, Cuba, Egypt, Ethiopia, Iran, Iraq, Israel, Libya, Mongolia, North Korea, Somalia, South Yemen, Syria, Vietnam.
Developed: 1957-1960
Manufactured: 1960-1972
Crew: 4 men
Fighting weight: 38.0 tons
Overall length: 9335 mm
Hull length: 6630 mm
Width: 3270 mm
Height: 2395 mm
Armament: 1 115 mm U-5TS (2A20) tank gun (40 rounds)
           1 7.62 mm SGMT or PKT machine gun
           1 12.7 mm DSchK-M AA machine gun (500 rounds)
Powerplant: 12-cylinder W-2-62 WR Diesel engine (580 HP/427 kW, later 620 HP/456 kW, finally 680 HP/500 kW)
Top speed: 50 kph
Range: 400-600 km

As already mentioned, at the beginning of the sixties extensive attempts to increase the combat value of the T-62 took place. Along with the requirement for greater power from the Diesel engine, higher mobility was particularly requested. One of the greatest weaknesses of the T-54, T-55 and T-62 was in the torsion bars of the running gear. Those of the first and last road wheels in particular quickly broke under sudden pressure. To test a new type of running gear, at least two T-62 medium tanks (1960 model) were equipped with six medium-sized road wheels on each side. These two prototypes ran under the designations of "Object 167" and "Object 167 GTD." The first of these was fitted with the higher-performance 700 HP W-2K Diesel engine, and the second with a 760 HP gas turbine. The two prototypes were the first step in the

The T-62 (1960 model) with KMT-5 ("roller section") mine removal device.

development, lasting to the early seventies, of a completely newly conceived modern medium tank, the T-72. At the same time, there began the development of another medium tank, which eventually resulted in the T-64. This apparently parallel development, though, was an inseparable process, which may be unique in history, and which cannot be understood from a purely technical viewpoint, as has been the case until now.

As of 1965, the T-62 was also fitted with a 12.7 mm DSchK AA machine gun (1938/46 model). This T-62 medium tank (1965 model) is seen during chemical training of the "Marshal of the Armored Troops Michail Efimovich Katukov" Guard Regiment.

T-62 (1965 model) tanks of the South Yemen Army on parade.

Winter action of a T-62 (1960 model) with BTU-55 (1963 model) dozer. The panel ahead of the driver was part of the equipment of all T-54, 55 and 62 medium tanks and was called a "driver shield."

The improvements made to the T-54/55 were also made to the T-62. These T-62M medium tanks are seen in a field parade in Afghanistan.

Chapter 2: The Russian Armored Vehicles of World War II

To test improved running gear and an upgraded powerplant, two prototypes, "Object 167 GTD" (tactical no. 219) and "Object 167" (tactical no. 218) were built on the basis of the T-62.

The T-62 was also fitted with a laser range finder above the tank gun. It was designated Medium Tank T-62M (1976 model).

The Medium Tank T-62M (1980 model) with laser range finder and added armor on both forward sides of the turret.

3

The Path to the Modern Battle Tank

By the end of the sixties, two different types of medium tanks—as seen from outside—had been created for the Soviet forces. Even though, to this day, only a partial view into the Russian archives has been allowed, and not all the background information on the development is known at this time, the facts are sufficient to explain this process.

Of the various theories that were originated, most were speculation. The high point may have been the statement of one author that the two vehicles came into being to encourage competition between two design offices. Economic factors, though, can be ruled out. They would have been foreign to a totalitarian state, and in crass contrast to a planned economy. In all possible exceptions that could arise under special conditions, often depending on individual personalities, the armaments industry in a state in which the political and military leadership are one and the same power, and the planned economy is regarded as practically sacred, has very little room for diversion.

Instead, the fact that for more than twenty years various models of medium tanks have been used by the armed forces of the USSR or the CIS/Russia may be the result of a complex process. Along with the political changes through the "choice" of Nikita Khrushchev to be Chairman of the Supreme Soviet and Secretary-general of the Communist Party (he is said to have had a particular fondness for rockets), it was primarily the concepts of the military theorists about the waging of war under the conditions of the use of atomic weapons. If the armored divisions of the army corps had been placed in the second echelon until then and were only to come into action when the motorized rifle divisions had overcome the enemy's main defensive lines, in order to penetrate deep into the enemy's defenses, the massive introduction of nuclear weapons brought about plans to use the armored divisions in the first echelon. The main task of the armored divisions in an attack with nuclear weapons was now to be breaking through, surrounding and passing enemy forces. The breakthrough was to break up the enemy's defenses, separate enemy troop units from each other, and create breaches for the second echelon to enter. The use of so-called "penetration regiments" thus became superfluous, and these were also restructured into typical armored regiments. Surrounding the enemy was to begin with a thrust into the open flank. The main troop groupings were to be circumvented.[14] But this required a kind of "universal tank," which simultaneously united in itself the characteristics of light, medium and heavy tanks. It was to attain strong armament and high speed, have a long range and have adequate protection against nuclear radiation.

The available light tanks were also used for armored reconnaissance, but the heavy tanks would no longer fit into this concept. The firepower of the tanks could only be attained through a quicker succession of shots and faster targeting. Their speed suffered from their great weight and the resulting burdens on the power train and running gear. The required amount of steel could no longer be justified, as the hollow-charge shells introduced for most types of weapons could penetrate all known types of armor. The relatively slow-moving heavy tanks could easily be defeated by anti-tank guided missiles.

The required armor protection could be attained only by using alternative armor types. Experimentation with multilayered armor was making great progress. Surface-hardened light metal alloys with ceramic inlays hindered the penetration of hollow-charge shells. Low-caliber shells should bounce off the very sharply angled armor plate. In order to keep the silhouette as low as possible, the height of the drive train, particularly the powerplant, had to be reduced. The W-2 Diesel engine used until now could not be made any flatter. A completely new engine had to be built. Protection from radiation had to be gained by attaching lead-asbestos plates in the interior. A higher rate of fire could be attained only by using an automatic loading system. The greatest source of errors in targeting was the determination of distance by means of a graduated plate in the aiming scope. The running gear had to even out the greater part of the uneven terrain at high speeds. But this could be achieved only by light running gear.

All of these requirements were not new, and many of the suggested solutions were known in theory for a long time, both nationally and internationally. Even the armored aprons used since the twenties were a form of multilayered armor. Automatic loading had been used by the navy for decades. Light metals with high stability had been used in aviation since the thirties. The Christie models were already fast-moving tanks. Range-finding devices already existed before World War I. The list could be extended at will. The

Chapter 3: The Path to the Modern Battle Tank

real innovation between the mid-fifties and early sixties, though, was the fact that science and technology had made significant progress in their developments. It was the time when electronic data processing was born, the atom was split, atomic machines and generators appeared, the first flights into space were made, intercontinental rockets were built, and the depths of the seas and the north and south poles were explored. In technology, there seemed to be no limits to development. In addition, the economy of the USSR had meanwhile recovered from the results of World War II. The arming of allied armies with Russian weapon technology brought in additional financial means. The reparations that had to be paid to the USSR as a result of the surrender of the German Wehrmacht, and which were paid exclusively by East Germany, added up to 2.2 billion Eastern Marks. In addition, 6.5 per cent of East Germany's national income was devoted to maintain the forces stationed in that country. The occupation troops made no further *de facto* expenses for the Soviet government.[15] In addition, 202 former armaments industries in the Soviet occupation zone had been dismantled and sent to Russia.

Numerous East German businesses of the Soviet Stock Company (SAG) supplied raw materials or produced exclusively for the USSR.[16]

The government of the country, which consisted for the most part of generals and officers of the armed forces, put no financial limits on the armaments industry. The reduction of the Soviet Army by 1.2 million men in February 1956 saved a considerable amount. At the same time, numerous international disagreements escalated into armed conflicts. The high points were the Cuban crisis, the Berlin conflict and the Vietnam War. It seemed clear to every Soviet citizen that the military had to have absolute priority, and in any case no opposition was allowed.

The end of heavy tank construction freed important production facilities for the building of other tank types. At the same time, large numbers of highly qualified technicians, engineers and scientists became available at one stroke.

The T-64 Medium Tank

For the development and production of the required new tank types, a completely reworked production line had to be created. The choice was made of Locomotive Factory No. 183 in Kharkov. The reason for it was surely the decentralization of industry decided on by Khrushchev in 1957 despite opposition from the economic experts. An important influence on the decision to include the tradition-rich factory in Kharkov in the development of new battle tanks may also have been the fact that production of the T-54 had begun with the Polish firm Kum BUMAR in Labendy and the Czech firm ZTS in Dubnica in 1954. This was meant to supply not only their own armies with T-54 tanks, but also the other Warsaw Pact countries. Production of the T-54, the successor model of which, the T-55 (1958 model) was already in series production in Nishni Tagil, could thus be reduced. A change or modernization of the production facilities was necessary in any case. In 1959, for example, the East German People's Army received the first thirty T-54 tanks from Poland. Until then, the East German government could only buy used T-54 tanks from the Soviet Union. The first 210 T-54s in East German service had been mustered out by the Soviet forces, and the price paid by East Germany had no relation to the technical condition of the tanks.

The direction of the development branch of the Kharkov factory was turned over in 1958 to Alexander A. Morosov, who was transferred, along with several members of his design bureau, from Nishni Tagil in the Urals to Kharkov in the Ukraine. They had designed the T-54 and had the most experience in the development of medium tanks. Thus, it can be assumed that A. A. Morosov had great personal influence on this decision, for he was born in Kharkov, and had had to move his "Morosov Construction Bureau," where the plans for the T-34 originated, from Kharkov to Nishni Tagil in 1941 as the front approached.

In Nishni Tagil, Leonid N. Kartzev had meanwhile taken over the developmental department of the tank factory, now known as the "Ural Wagon Design Bureau." It is certain that Kartzev was given the office of the new director with Morosov's recommendation, and a close personal relationship probably existed between Morosov and Kartzev in the years that followed.

While Morosov, with his staff, was to develop a completely new medium tank, Kartzev and his people had the job of improving the existing tanks. All the new knowledge gained in tank development was made available to both offices.

While Morosov and his staff looked for ways to design a tank that could meet all its requirements, the Ural Design Bureau continued its efforts to improve tank armaments. By the mid-fifties, a smooth-bore 100 mm tank gun had been developed for the T-54 medium tank. The research project went by the concealing name of "Object 140." With this prototype, important knowledge of the ballistic performance of smooth-bore guns in battle tanks was gained. Considerably better firing performance was attained by increasing the caliber to 115 mm. In 1959 the prototypes, designated "Object 165" and "Object 166," were finally fitted with 115 mm smooth-bore tank guns. One of the two prototypes then became the T-62 Medium Tank (1960 model), which required a new turret shape and minor modifications on account of its heavier armament. The other may have been the standard T-54/55 with the 115 mm smooth-bore tank gun.

Parallel to this, the prototype of the new medium tank had come into being at Kharkov by 1960. This developmental model was designated "Object 430." At least two prototypes were produced. One of them was fitted with the 100 mm smooth-bore gun, the other with the 115 mm smooth-bore type. Finally, the D-68 tank gun, already used in the T-62, was introduced as U-5TS (2A20) with a caliber of 115 mm. So as to be able to fulfill the requirement for a low height of the armored hull, the principle, already developed in 1894-95 by the Berlin engineer Nadrowki for a combine motor,[17] was taken up and an opposed-piston motor with two horizontal rows of cylinders was built, the 5TDF, which attained a performance of some 760 horsepower.

---
14. Shilin, P. A., Istorija woennogo iskusstwa, Moscow 1979, p. 26.
15. Heinze, S., Studie zu Charakter, Aufgaben, Struktur, Bewaffnung, Ausrüstung und Entwicklung der KVP, Dresden 1990, p. 17.
16. Hoffmann, J., Ein neues Deutschland soll es sein, Berlin 1989, p. 49.
17. Sittauer, H. S., Diesel-Eine Erfindung erobert sich die Welt, Leipzig 1961, p. 96.

Overrevving of the powerful engine was prevented by an RPM governor, such as was used on all Russian tank engines. This limited the speed of the tank to a speed between 65 and 70 kph. The chief visible feature was now the development of suitable running gear for the tanks. It was known that only lightweight road wheels could even out bumps in the ground at high speeds. Small road wheels also allow a large spring path and lighter or shorter torsion bars. The decision was made in favor of short torsion bars, which could now be removed from inside the hull and mounted over the road wheels. This resulted in a further decrease of the hull height. This principle was soon discarded because of the unprotected mounting of the road wheels. Tracks running on return rollers took more pressure off the road wheels. The greater the number of road wheels, the better is the weight distribution on the surface of the tracks. This in turn allows, with a constant weight, the use of narrower tracks, the weight of which is again less. These were experiences that were contributed by engineers and technicians and had formerly been used in the development of heavy tanks. Through the use of a combination of earlier armor steel, light-metal alloys and ceramic inlays it was possible not to exceed the weight of the T-54/55/62 and, also using diagonally placed armor plates, to attain greater armor protection.

While tank production at Nishni Tagil was converted to the manufacture of the T-62 Medium Tank (1960 model), experiments were begun to utilize the results of chassis testing with "Object 430" (1960 model) and "Object 167" (1961 model) to improve the T-62. In addition, tests were begun with "Object 167 GTD" (1962 model) to increase engine power by installing a gas turbine.

Parallel to the development of a new motor and an improved power train, the Morosov group concentrated on increasing the rate of fire and creating a new targeting system. In 1963, "Object 432" (1963 model) was introduced; externally it was like "Object 430" (1960 model), but it had an automatic loading system and a TPD-43B optical basic range finder. With the automatic loader, the theoretical rate of fire of the tank gun could be raised from four rounds per minute (one to two shots per minute in practice) in the T-62 with hand loading and a loader dispensed with. The ammunition feed principle, which was designated "Kosina" (basket), had thirty shells, in angled containers, attached to the turntable around the tank gun. The aiming gunner could electrically position the turntable so that the container with the right kind of ammunition was positioned behind the breech. Then the "basket" was lifted and a shell pushed into the tank gun. After firing, the empty cartridge case was ejected and fell back into the container.

With the principle of overlaying the pictures of a target by using two optically separate scopes, the range could be determined considerably more accurately and the gun's angle be set more precisely.

By 1964, testing of the "Objects 167, 167 GTD and 432" at the test center in Kubinka was finished. The "Object 432" prototype was so convincing that production of this tank began in Kharkov in the same year, and one year later at Nishni Tagil, under the designation of Medium Tank T-64 (1963 model). Within three years, over 600 of them had been completed.

Meanwhile, tests were being conducted to improve the probability of hitting targets at ranges over 1,000 meters. Through the new targeting system, enemy armored vehicles could be spotted at a distance of about 2,000 meters. The range of direct shots from the 115 mm U-5TS tank gun, though, was only about 900 meters for the low-caliber shell and 600 meters for the hollow-charge shell. At the same time, the development of antitank guided missiles showed great progress. Numerous vehicles equipped with antitank missiles had already been introduced. Since 1959, numerous experiments with the XM 81 cannon rocket launcher of the XM 551 light tank had been carried out to make it possible to fire missiles from the tank gun. Thus, it was important to utilize the results of this development in the realm of antitank missiles for tanks as well. The arming of "Object 167 (1961 model)" with an externally attached launcher for the 9M14 "Malyutka" antitank missile seemed to be one solution, but had the definite disadvantage of being unprotected from shots and splinters. Using the rockets under nuclear-weapon conditions was also impossible. As an alternative, the design bureau of Pavel P. Isakov, at the Chelyabinsk Tractor Works (Factory no. 200), which had occupied itself with the development of heavy tanks since 1960, suggested that a recoil-free gun be installed in a tank turret, so as to be able to fire guided and unguided missiles. At the same time, the striving for as low a silhouette as possible was completed with the relocation of the driver's seat and steering controls to the turret. The driver's function was to be assumed by the commander. An automatic loading system made a loading gunner unnecessary. Thus, the tank's crew consisted of only two members, the commander-driver and the aimer-gunner. The tests took place under the concealing designation of "Object 775 (1962 model)." The ammunition load consisted of 24 hollow-charge guided missiles and 48 rocket shells with splintering and explosive effect.

From the design bureau of the Kirov Works in Leningrad, directed by S. J. Kotin and likewise included in the development and construction of heavy tanks, came a suggestion for a rocket carrier. Along with the two automatically loaded recoilless 73 mm guns on the front, each with 32 rounds of ammunition, there was a folding rocket launcher with fifteen launching rails on the rear of the tank. The crew consisted of four men: the commander, driver, and aiming gunners 1 and 2. The vehicle went by the designation of "Ob-

Forerunner of the T-64: the "Object 430" medium tank (1960 model), armed with a 100 mm tank gun.

Chapter 3: The Path to the Modern Battle Tank

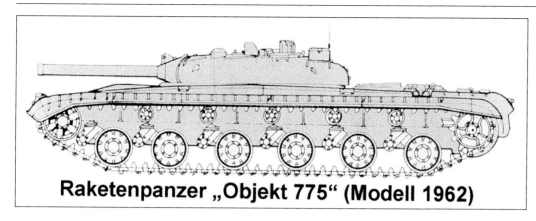

The "Object 775" rocket tank (1962 model) could launch guided missiles and guided artillery ammunition with its launcher.

ject 287" (1962 model). To create an armored vehicle with an extremely flat silhouette and especially heavy armor, "Object" 288" (1962 model) was built. Although the projects were not taken into consideration for the development of a new medium tank, important elements were taken over for the BMP-1 armored troop carrier developed in the mid-sixties.

Thus, there were three different types of medium tanks produced simultaneously in the USSR in the sixties. Production of the T-55, T-62 and T-64 took place in Kharkov, Omsk and Nishni Tagil.

All the requirements of a modern battle tank appeared to have been attained in the T-64 medium tank. It was generally regarded as reliable, meaning that it suffered technical problems no more or less often than the forerunner types. There are occasional references in the literature to the motor not being a fully developed design and showing a large loss of power when overtaxed, to a large loss of speed resulting from taking turns too sharply, to the loading system being unreliable, and to the fighting compartment being so cramped that loose parts of uniforms could get stuck. These "faults" were found in all Russian tanks from the T-54 to the T-72 and were not particular characteristics of the T-64.[18]

The automatic loading system, though, brought about one disadvantage: during the loading process, the tank gun had to take a certain position and the aiming mechanism was without power. For a certain time, aiming was impossible. In order to shorten the loading time, A. Schomin suggested a way to shorten the ammunition feed. This was done, on the one hand, by separating the cartridge from the shell, so that the shell nearest to the loading arm was loaded, and on the other, making the shell a larger caliber to maintain its effect so its length could be shortened. The ammunition, now lying in star formation under the turntable of the gun in two separate ammunition carousels (the cartridges above, the shells below), now allowed the carousels to turn at a rate of 70 degrees per second, shortening the loading time by six to eight seconds. The T-64, now fitted with the 125 mm smooth-bore gun, was tested as "Object 434" (1965 model) and was introduced as Medium Tank T-64A (1965 model). Internationally it was first seen in the southern group of the Soviet fighting forces in Hungary. In the time that followed, it was introduced particularly among the armored units stationed outside the USSR. For example, it was used by the 2nd and 3rd Stoss Armies and the 2nd and 20th Guard Armies.

Numerous improvements were also made to the T-64A, more than 8,000 of which were produced in Kharkov until 1975, and 3,997 of which were still said to be on hand in 1991. Along with its equipping with a second radio set and a nuclear warning device, which generally were given the suffix "K," as of 1974 the TPD-7-49 range finder was replaced by a TPD-K-1 laser-scope range finder. Sighting the target just required one more sighting aperture, which was located on the left front side of the turret roof. In the T-64A, which was later reequipped with the new targeting system, the right opening was usually closed with a steel plate.

Development of the battle tank of the eighties was already concluded at Kharkov in 1976. As always, this development was not broken off, but modified steadily. The last series of the T-64 took on all the improvements of the new tank, which was introduced as the T-80. The 9M117 rocket of the 9K112 "Cobra" guided weapons complex (NATO code AT-8/SONGSTER) had been developed for it. In addition to the installation of the improved 125 mm D-81TM (2A46) tank gun, a new sighting block, from which the commander could aim the rocket, had been mounted in front of the commander's cupola. Some 1,200 of this tank model, designated T-64B (1976 model), had been built by 1982.

Parallel to the developmental work on the 9K116 "Bastion" guided weapons complex, which included the 9M117 rocket for the 100 mm tank gun, the 9K120 "Swir" guided weapons complex had been developed for the 125 mm tank gun (2A26) in tanks not equipped with the sighting block. In this system, the 9M119 rocket was guided by a laser beam. The 1K13 targeting device served as the targeting optics and was installed near the aiming gunner in place of the night-vision targeting scope. It can be assumed from this that these T-64 tanks were suffixed with "B1" in their designation.

As of 1983 the T-64 tanks were fitted with a gas turbine with its power raised to 1,000 HP (735 kW) in addition to the already introduced modernizations when they were given major service, under the designation of T-64BM medium tank.

Between 1982 and 1984, experiments were conducted to protect the tanks better from hollow-charge shells by using the already internationally known principle of reactive armor. The development was directed by W. N. Brishkov of Department N II of the Research Institute for Battle Tanks in Kubinka. In 1985 the introduc-

---

18. The author himself was for many years an instructor on almost all Russian armored vehicles, and is presenting here his personal experiences with Russian tank technology.

From 1963 to 1969, the T-64 medium tank (1962 model) was built with the 115 mm D-68 (U-5TS) tank gun of the T-62. In these tanks the infra-red firing spotlight was still mounted to the right, near the tank gun. As of the eighties, the first version of this type was probably designated T-64R.

A platoon of a guard armored division of the Soviet forces in Germany, equipped with T-64A medium tanks (1965 model).

tion of this "additional armor" finally began, and all tanks that left the factories with it or were updated with it were given the suffix "W."

In the same time period, the turret roofs of the T-64 were given additional armor for better protection against "Top-attack" ammunition fired from aircraft or by indirect aiming.

Command tanks of all T-64 versions were also produced, and their designations also had the letter "K" added. The commanders of the tank battalions had a second radio set, a navigational system, a nuclear warning device, and a semi-telescopic mast. The vehicles of company chiefs were equipped with a second radio set. For reasons of space, the ammunition capacity had to be reduced in both versions. The vehicles of battalion commanders also went without the 12.7 mm NSWT anti-aircraft machine gun.

Medium Tank T-64A, T-64B
Sredniy tank T-64A (obr. 1965 g.), T-64B (obr. 1976 g.)

Made in: USSR (tank factories in Kharkov and Nishni Tagil)
Used in: USSR/CIS/Russia, Ukraine
Developed: 1965
Manufactured: 1967-1982
Crew: 3 men
Fighting weight: 38.0 tons
Overall length: 9225 mm
Hull length: 6400 mm
Width: 3380 mm
Height: 2170 mm
Armament: 1 125 mm (2A26) or D-81TM (2A46) tank gun (40 rounds)
　　　　　1 7.62 mm PKT machine gun (3000 rounds)
　　　　　1 12.7 mm NSWT anti-aircraft machine gun (500 rounds)
Powerplant: 10-cylinder 5TDF opposed-piston engine (760 HP/559 kW)
Top speed: 70 kph
Range: 450 km

With the introduction of the 125 mm 2A26 tank gun and the inclusion of a TPD 7-49 base-tube range finder, the T-64A medium tank (1965 model) came about. It was developed from "Object 434" (1965 model) and was built from 1967 to 1969.

Chapter 3: The Path to the Modern Battle Tank

The tanks of this production run could also be fitted with the KMT-4 mine removal device. Shown here is a T-64A (1965 model), along with a BMP-1 infantry combat vehicle (1966 model).

With the introduction of the improved 125 mm D-81TM (2A26) tank gun and the 9K112 "Cobra" guided missile complex, the T-64B medium tank (1976 model) was formed. A sight block with which the commander could direct the 9M117 rockets was installed. For the T-64, which did not have this block installed, the 9K120 "Swir" guided missile complex was developed. In these tanks, which were also known by the designation T-64B1, the night targeting scope could be replaced by the 1K13 targeting device to guide the 9M119 rockets. The picture shows a T-64BW (1976 model) with sight block, plus holding bolts for the reactive armor plate.

In the mid-eighties, the turret roofs were fitted with added protection against aircraft and ammunition fired by indirect aiming ("top attack"). This T-64B medium tank has an enlarged opening for the 1K13 targeting device and is listed in KSE documents as T-64B1.

A T-64BW medium tank fitted with added reactive armor plate.

As of 1967, the base-tube range finder was replaced by the TPD-K-1 laser targeting scope. In the vehicles built from 1967 to 1976, there was just one vision port for this device. The T-64, which was reequipped with the new scope, had the right opening for the base-tube range finder covered with a steel plate.

The T-72 Battle Tank

While the production of the T-64 was begun in 1964-65, the Ural Wagon Design Bureau in Nishni Tagil was occupied with experiments aimed at producing the new tank with traditional rolled and cast steel armor. This was surely done from the standpoint that the new tanks would also be produced for Warsaw Pact adoption in Czechoslovakia and Poland, where supplies of aluminum were not available. Independently of that, the tank of the next generation appeared on the drawing boards of the Kharkov tank works under the designation of "Object 219."

The first prototype of the "T-64" with old-style steel armor plate, designated "Object 172," was finished in the summer of 1968. Since it was seen in its first tests in the country that the total weight, some five tons higher, put too much pressure on the running gear, the chassis tested on "Object 167" was used for the "Ural Tank."

For reasons of simplifying production and maintenance, the first prototype already had a 12-cylinder Diesel engine installed. This may have been primarily a result of decisions made by the Warsaw Pact governments, for this motor and its components were used in all the Russian battle tanks and self-propelled guns and produced in all the pact nations. The Chelyabinsk Motor Design Bureau was entrusted with attaining the maximum increase in engine power.

At the end of 1968 the first prototype, with the chassis of "Object 167," was completed as "Object 172 M." By the summer of 1969, more of these tanks had been built and were being tested by the troops in central Asia. The tests continued until 1971, taking place north of China and in the trans-Balkan area, among other places. In 1972 the equipping of the first tank regiment with these tanks, now designated T-72, finally began. After a successful conclusion of the two-year troop testing, the introduction of the T-72 began in 1972. Now not only the tank factory in Nishni Tagil, but also those in Chelyabinsk and at the Kirov Works in Leningrad were included in its production. The first series of the T-72, like the T-64/T-64 A, was equipped with the TPD-7-49 optical range finder. While the prototype and the pre-series vehicles had used the L-2AG infra-red firing spotlight, like that of the old T-64, attached to

This vehicle, from the first T-72 (1971 model) production run, grew out of "Object 172M" (1968 model) and had the running gear of "Object 167." The L-2AG light was still attached to the left of the 125 mm tank gun.

the left side, it was now attached to the right side alongside the tank gun.

When the introduction of the TPD-K-1 laser range finder began in 1974, the T-72 tanks that were equipped with optical range finders were also reequipped. Here, too, the unneeded opening on the right side was finally covered with a steel plate. The tanks being delivered with the laser range finder were designated T-72A ("Object 174") by the Soviet Army. In the Warsaw Pact countries they were known as T-72M.

As of 1976, the 9M119 tank-gun rocket of the 9K120 "Swir" guided weapons complex was included in the armament of the T-72. Here, too, the 1K13 targeting device was installed instead of the night targeting scope.

In 1979 the 902A fog-grenade firing device was added, the improved TWNE-4B night targeting device was installed for the driver, the ammunition was increased from 39 to 44 rounds, a W-46-6 Diesel engine upgraded to 840 horsepower was used, and several smaller improvements were included. In addition, the turret front was strengthened, an essential change.

In 1982 the armor on the front of the hull was increased by 16 mm, and the turret was given convexities on its front in which ceramic tiles were set as additional protection against hollow-charge ammunition. These clearly noticeable rounded protuberances gave the turret the name of "Dolly Parton" in the West, after the country singer who was not renowned for her voice alone. Officially the modernized "Object 174" vehicle was introduced as the T-72B medium tank. Since the leadership of the Warsaw Pact nations had already introduced the T-72A as T-72M, the new version was known as T-72M1.

In 1985 the turret front, turret roof armor and driver's front area were strengthened again. The tanks, known as T-72B1 among the CIS/Russia armed forces, were called T-72M1M ("Super Dolly Parton") by the international press. The additional armor plate on the turret, some 25 mm thick, was made of plastic strengthened with fiberglass. It was intended primarily to protect against fire from "intelligent" ("Top-Attack") ammunition. In addition, the 902A foglaying apparatus was moved farther toward the back of the turret, so the front could be covered with a reactive additional armor.

With "Object 172" (1968 model) there arose a medium tank that resembled the T-64 externally; but its hull was made of old-type armor steel and it had a 12-cylinder, 780 HP Diesel engine.

Chapter 3: The Path to the Modern Battle Tank

Rohrrakete 9M119 der 125-mm-Granate mit Ausstoßladung 9CH949 des Lenkraketenkomplexes 9K120 „Swir"

The ammunition of the 125 mm 2A26 tank gun (from left to right): splinter explosive shell OF, hollow-charge shell BR, undercaliber shell K with partial charge, undercaliber shell without partial charge, cartridge ignition charge, transport rack for cartridge ignition charges.

As early as 1984 the tests of reactive auxiliary armor were concluded. One year later, the installation of brick-shaped explosive segments on the armor-plate bolts began. Some of the T-72 tanks turned over to the troops were equipped with attachments during their major refitting at the plant where they had been built. Although no documentation of their official designation has been found to date, it can be assumed that they were called T-72BW or T-72B1W. Internationally, though, the designation T-72S has been used.

Around 1988-89 the installation of the 9K119 "Reflex" guided missile apparatus in the T-72 began. The new 1G46 "Irtysch" targeting device, which serves as the targeting scope for the 125 mm (2A46M1) tank gun and the targeting optic for the laser-beam-guided 3-UBK-20 rocket, had been developed for the T-80. Here the driver's area and the front track aprons of the T-80BW were adopted, as was the attachment of reactive armor plate on the turret front. In the Russian forces these tanks were known as T-72BM.

Between 1992 and 1993, all the innovations in the realm of tank technology were at last included, in the customary way, in the continuing production of the T-72 at the Nishni Tagil tank works. The goal of this modernization was not only increased combat value, but particularly the pressure of meeting the competition in a free-market economy. The turret and hull were constructed in multilayered fashion, and the armor plate was covered on the outside with reactive "Contact-5" auxiliary armor. In the turret, the same computer-supported fire control system with laser range finder as is used in the T-80U was installed. A row of sensors were attached to the turret to give the calculator ballistic values. The installed 125 mm D-81TM (2A46A) "Papira-3" tank gun was able to fire the laser-directed 9M119 "Reflex" rockets. The range of the projectiles is 5,000 meters. Thanks to the laser-beam direction, the rockets can now be fired while the tank is moving. The "Agawe" heat sensor can also be installed for targeting use. Through the use of night vision apparatus, the usual firing spotlight can be eliminated. In addition, the tank, now designated T-90, has a laser warning system and the TSchU-1-7 "Shtora" protective system to deceive approaching antitank missiles. For this purpose, powerful infrared lights are attached on either side of the tank gun. The protective system is activated when a warning sensor has become aware of an enemy laser beam. Immediately the turret turns in the direction of

The T-72 medium tank (1971 model) was still equipped with the TPD-2-49 base-tube range finder.

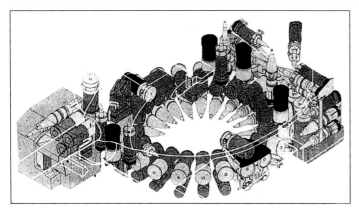

Diagram of the intricate weapons loading system.

When driving on land, the air duct was attached to the left side of the turret.

the laser-directed projectile or laser range finder. At the same time, both infrared lights send signals to the enemy aiming controls that will disturb the course of the projectile. The launching of foglaying shells in the direction of the enemy line of fire not only provides camouflage but also interrupts the enemy guiding beams.

In terms of its chassis, the T-90 corresponds thoroughly with the T-72BM.

The T-72 commands tank had the suffix "K" added to its designation. The vehicle, intended for the commanders of tank battalions, carried combat equipment that included 33 rounds of ammunition. In addition to a second (R-130M) radio set, a TNA-3 navigation device and a gasoline-powered AB 1p/30-M generator were firmly installed. In addition, a semi-telescopic mast ten meters long was attached at the rear of the vehicle. The company chief's tank carried 39 rounds and had a second R-123M radio set installed.

Above and Right: For underwater driving, the air duct was set into the signal opening of the aiming gunner's port.

Chapter 3: The Path to the Modern Battle Tank

It has often been said that the folding side panels were supposed to give increased protection from hollow charges. On the other hand, they were set up during attacks, offered no lateral protection and covered no surface from the front. They may have been adopted simply because they fooled enemy range finders into seeing them as extensions of the tank.

Medium Tank T-72, T-72A, T-72B, T-72B1, T-72BM, T-90
Sredniy tank T-72 (obr. 1972 g.), T-72A (obr. 1974 g.), T-72B (obr. 1982 g.), T-72B1 (obr. 1985 g.), T-72BM, T-90.

Made in: USSR (tank factories in Nishni Tagil, Leningrad, Cheljabinsk), CSSR (Dubnica, Novy Jicin, Pilsen, only T72A/B), Poland (Labedy, Katowitze, only T-72A/B), India (Avadi, only T-72A/B), Iraq, Yugoslavia
Used in: USSR/CIS/Russia, Ukraine, exported: only T-72/T-72A/T-72B, to Algeria, Bulgaria, CSSR, Cuba, East Germany, Egypt, Ethiopia, Finland, India, Iraq, Kuweit, Lybia, Madagascar, Poland, Romania, Saudi Arabia, Syria, Yugoslavia

	T-72	T-72A	T-72B	T-72B1	T-72BM	T-90
Developed:	1971	1974	1981-82	1985	1988-89	1991-92
Produced:	1972-74	1974-82	1982-85	1985-90	1989-95	1993-
Crew: 3 men						
Fighting weight (t):	41.0	41.5	43.0	44.5	44.5	46.5
Overall length (mm):	9530	9530	9530	9240	9240	9654
Hull length (mm):	6860	6860	6860	6570	6570	6910
Width (mm):	3460	3370	3370	3370	3590	3590
Height (mm):	2370	2430	2370	2370	2370	2682
Armament:	1 125 mm D-81TM tank gun					
	2A46	2A46	2A46	2A46	2A46A	2A46A
Gun shells:	39	44	44	46	45	45
	1 7.62 mm MG PKT, PKT-M					
Rounds:	2000	2000	2000	2000	1000	1000
	1 12.7 mm Fla-MG NSWT					
Rounds:	300	300	300	300	300	300
Fog cartridges:	0	12	12	8	8	6
Powerplant:	12-cylinder multifuel engine with mechanical charger					
	W-46	W-46-6	W-46-6	W-84	W-84	W-84A
Power (HP/kW):	780/573	792/582	792/582	840/618	840/618	842/619
Top speed (kph):	60	70	70	70	70	65
Range (km):	650	600	600	550	550	650

Internationally used designations for T-72 and successors, including modernizations

T-72 = T-84 (Yugoslavia), TR-125 (Romania)
T-72A = T-72M (Warsaw Pact and export)
T-72B = T-72M1 (Warsaw Pact and export), "Assad Babil" ("Lion of Babylon", Iraq, T-72M1),
M-84 and M-84A (Yugoslavia, modernized T-72M1), T-72M2 (Czech Republic, modernized T-72M1), PT-91 TWARDY (Poland, with reactive armor, 1994), T-72M2 MODERNA (Slovakia, modernized T-72M1), "Dolly Parton" (USA)
T-72B1 = T-72M2 "Super Dolly Parton" (USA)
T-72BW (?) = T-72S (international specialist press)
T-72BM = SMT M1990 (international specialist press)
T-90 = "Follow-on Soviet Tank No. 1" (USA or "Future Soviet Tank" (NATO)

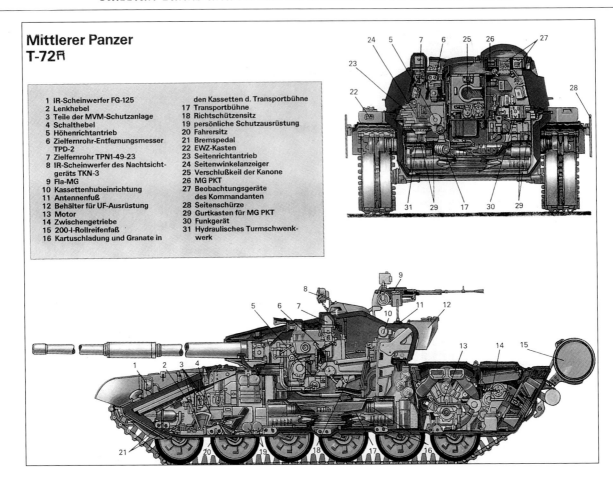

In the East German People's Army and other Warsaw Pact forces, the T-72A (1974 model) was called T-72M.

Chapter 3: The Path to the Modern Battle Tank

To remove mine barrages, the T-72 medium tank could be fitted with the KMT-6 or KMT-6M mine removal device ("knife section"). Compared to the KMT-4, the KMT-6 was wider, and the KMT-6M still had a folding side panel.

After the German reunification, the Bundeswehr of the Federal Republic took over 551 T-72 medium tanks on October 3, 1990. Altrhough none was used, a few reached the scrapyard under their own power.

At the beginning of the eighties, new side aprons and the 902A fog-cartridge firing device were introduced. This T-72A medium tank (NVA: T-72M), which is in the Military History Museum of Dresden today, was the first tank of this production run that was delivered to East Germany by the USSR. Until 1990 it was in use at the Military Academy in Dresden.

The optics of the TDK-K-1 laser range finder and the splinter shield over the TPN-1-49-23 infra-red night targeting scope of the T-72A medium tank.

The author driving a T-72A in 1984. Under the bow, the plow blade, 2140 mm wide, can be seen. With it, cover for the tank could be created in 12 to 40 minutes, depending on the type of ground.

The launchers for the 902A fog-cartridge apparatus on the right side of the T-72A's turret.

Between 1980 and 1982, ceramic tiles were set into the front of the turret on both sides of the tank gun. This gave the medium tank, now designated T-72B (1982 model), the nickname "Dolly Parton" in the NATO forces. In the Warsaw Pact states it was called T-72M1.

The T-72B1 medium tank (1985 model) of the "Yuri Andropov" Guard Armored Division, seen on parade in Moscow in November 1986.

The T-72B medium tank, with the reinforced turret front clearly visible.

In the T-72B1 medium tank (1985 model)—internatinally known as T-72M1M (or "Super Dolly Parton"), the turret and hull fronts were reinforced again. The turret roof also was given added protection against "top-attack" ammunition.

In the mid-eighties, medium tanks began to be given an added reactive armor. This included the T-72. Although the designation T-72S is often found in the international literature, the tank may have been called T-72BW or T-72B1M.

The T-72B1W medium tank with bolts for attaching the reactive armor, seen at the Guard Officer School at Ulyansk on the Volga.

Chapter 3: The Path to the Modern Battle Tank

The T-72BW medium tank, already equipped with the forward track panels of the T-80.

In the course of introducing the T-80, numerous improvements from the new tank were applied to the T-72 as well. The T-72BM medium tank had a new front design, different attachment of reactive armor on the turret front, and new front track panels.

The T-90 medium tank—undoubtedly an improved T-72.

The T-90 medium tank at the IDEX '97 in Abu Dhabi—the newest modernization of the T-72 from the tank works in Nishni Tagil.

4

The Modern Russian Battle Tank of the Guard Tank Units - the T-80 Medium Tank

For a long time the West wondered how the "Future Soviet Tank" of the USSR/CIS/Russia fighting forces in the eighties and nineties would look. As early as 1970 the designers at the Kharkov Tank Works had taken up the question of how the developmental results of the T-72 could be included in the production of the T-64. The research project was to be carried out under the camouflage name of "Object 219." Nikolai Shomin of the Morosov Design Bureau was entrusted with its leadership. Morosov himself was sent into retirement.

Particularly after the transition from light metal to multilayered armor had been made internationally, a considerable increase in weight could be expected. The T-72 had already become almost five tons heavier than the T-64 through the use of customary armor steel. To retain the T-64's off-road mobility and firepower as much as possible, the premise that only a high-performance powerplant and a flexible running gear could meet the desired prerequisites in a broad sense was surely the starting point. The design of a motor with at least 1,000 HP was to be taken on by the developmental office of the Kirov Works in Leningrad, directed by V. J. Klimov.

From the development contract to the first prototype, six years passed in all. The reasons for this may have involved the powerplant. That a single gas turbine of relatively small size can produce such extremely great power could have been taken for granted. It must have turned out to be complicated to retain the underwater running capability. This was a problem that all the international tank designers had to overcome in the seventies. When series production of the T-80B medium tank (1976 model), now equipped with the 9K112 "Cobra" guided missile system, could finally begin in 1976, it corresponded in ballistic technology with the T-64B and the T-72A. It had the same 125 mm D-81TM tank gun (2A46) and fired the same ammunition, but it differed essentially in its power-to-weight ratio. The 1,000 HP SM1000 gas turbine gave the tank a top speed of 70 kph. In loose sand, 50 kph could still be attained. The speeds of comparable heavy tanks at that time reached no more than 25 to 30 kph on loose ground. In connection with the high-performance running gear, the T-80 was and is capable of attaining very high speeds, even on very uneven terrain. In addition, the fire from the tank gun can be utilized on off-road terrain at medium speeds. The soldiers of the Soviet armored troops therefore gave it the nickname of "Flying Tank."

The numbers of T-80 and T-80B tanks actually built, though, were relatively small. The Soviet armaments industry built 266 tanks of these types, 75 in Kharkov, 13 in Omsk and 178 in Chelyabinsk.

The T-80 was introduced into the Soviet Army's guard units west of the Urals at the end of the seventies and in the early eighties. In 1984 the process of equipping the T-80 with reactive auxiliary armor and turning it over to the troops as the T-80BW was begun. As of the spring of 1984, the first T-80 and T-80BW tanks were turned over to the armored training units of the 1st Guard Tank Army, 2nd and 8th Guard Armies stationed in East Germany. But only at the end of July 1988 was it possible for Western observers to come face to face with this tank. And when the successor model, the T-80B with the 9K112 "COBRA" guided missile system, introduced in 1978, was shown, an "expert" in the specialist press could not conceal his disappointment. He wrote: "The pictures are unique but reveal nothing sensational . . . Except for the new type of gas-turbine drive, it is an evolutionary model, not a revolutionary new development."[19] It was the result of twelve years of service by the T-80 in the USSR's armed forces, that showed more of well-done concealment than of knowledge. A year before,

The result of requests to combine the advantages of the T-64 and T-72 medium tanks—the T-80 medium tank (1970 model).

---

19. Soldat & Technik No. 11, Frankfurt am Main, 1988, pp. 654-655.

Chapter 4: The Modern Russian Battle Tank of the Guard Tank Units

In 1976, after years of development, the T-80 was ready for production. With the inclusion of the 9K112 "Cobra" guided-missile complex, with a visor block ahead of the commander's cupola, the tank was designated T-80B (1976 model). Production went on until 1987 but included only 266 vehicles.

the Soviet Army had begun to introduce their newest battle tank. But it was introduced publicly only in Moscow during the parade on May 1, 1989. Meanwhile, the Russian government, for reasons of market economy, saw itself compelled to show this model at the IDEX 93 in Abu Dhabi. Officially it was offered as Medium Tank T-80U (U = "uluchennyi wariant"—"improved version"). Even though the T-80U was already six years old at this point in time, the three main points of tank evaluation (armor protection, firepower and mobility) can be seen as improved.

The armor plate on the front of the hull and the fronts of the side aprons was increased by means of added modular armor. This obviously consists of plastic in which ceramic pieces are contained. On the lower part of the hull front, two further rubber-weave aprons have been attached, intended to ignite mines with sharp bend ignition ahead of the vehicle. The basic shape of the turret corresponds to that of the T-72B1 designed in 1985, and accordingly also has "inset pockets" with ceramic tiles. In addition, V-shaped armor pieces have been attached to the hull front and the forward sides of the turret, meant to offer greater protection from hollow-charge ammunition with rubber-weave segments in the turning-circle area.

Added armor covers the roof of the turret from the tank-gun opening to the rear part. It is built up like that of the hull front. This should considerably decrease the effect of self-seeking ("Top-Attack") ammunition.

The firepower of the 125 mm D-81TM (2A46B, also known as 2A46M-1) tank gun of the T-80U was improved by the 1W528 electronic calculator with digital indication. After finding the target with the 1G46 "Irtish" targeting telescope-range finder, it indicates the angle of allowance according to the choice of ammunition. In the process, the shortening or lengthening of the range by the vehicle's own speed and that of the enemy tank are calculated through constant measurement by the laser range finder. In addition, the values measured by sensors, regarding the angle of the tank gun, barrel wear, outside and inside temperatures and air pressure, are considered. For firing at night, a remaining-light targeting scope, presumably with integrated heat sensitivity, is available. For firing the laser-beam-guided 9M117 rocket, though, this "BURAN-PA" night targeting device must be replaced by the 1K13 targeting scope-aiming device. Since this is stabilized at two levels, as is the tank gun, rockets can also be fired when the tank is in motion, but only in daylight. As with all Russian medium tanks since the sixties, the commander can override the traversing of the tank gun by the gunner. With the now-installed, vertically stabilized TKN-4S day-and-night observation and targeting device, targets can now be aimed and fired at by the commander.

With the installation of the 1250 HP GTD-1250 gas turbine, the T-80U attains the same power-to-weight ratio as the modern NATO tanks. Meanwhile, a 1500 HP gas turbine has also been developed by the Russian war materials industry. But it has also been said that since 1992 a 1000 HP 6TDF Diesel engine can also be installed. Presumably this 12-cylinder opposed-piston engine had been developed to give more power to the T-64 and is now available for the T-80U. The tanks that are equipped with Diesel engines are called T-80UD.

Some 6,000 T-80U tanks had been made by the time the USSR ceased to exist.

All the T-80 versions are also built as commanders' tanks. These have the usual "K" suffix to their designation and are equipped similarly to the T-72 command tanks.

Because of the high speed at which it could be driven, even off-road, the Soviet tank soldiers nicknamed the T-80 "the flying tank."

The medium tanks of the T-80 series could also be fitted with the KMT-6 or KMT-6M mine removal device.

Between 1984 and 1987, the T-80B was fitted with reactive armor and used by the USSR/CIS/Russia forces as the T-80BW medium tank.

Medium Tank T-80U
Sredniy tank T-80U (obr. 1987 g.)

Made in: USSR (tank factories in Kharkov, Chelyabinsk, and Omsk)
Used in: USSR/CIS/Russia, Ukraine
Developed: 1986-87
Manufactured: 1987 to present
Crew: 3 men
Fighting weight: 46.0 tons
Overall length: 9900 mm
Hull length: 7400 mm
Width: 3400 mm
Height: 2480 mm
Armament: 1 125 mm D-81TM (2A46B), (39 shells + 6 rockets)
   1 7.62 mm PKT-M machine gun (2000 rounds)
   1 12.7 mm NSWT anti-aircraft machine gun
Powerplant: GTD-1250 gas turbine (1250 HP/919 kW), GTD-1500 gas turbine (1500 HP/1104 kW), or 12-cylinder 6TDF Diesel engine (1000 HP/735 kW)
Top speed: 75 kph
Range: 335 km

Left and Right: **During the 1989 Moscow parade, the public was first shown the T-80U medium tank (1987 model).**

The T-80BW medium tank. *Below* **At first the T-80U was powered by the 1250 HP gas turbine. Since then, the Russian armament industry has offered a 1500 HP gas turbine and a 1000 HP Diesel engine.**

The latest modernization of the T-80 was presented by the Omsk factory in 1998. In addition to a new weapon rangefinder and an improved ammunition loading system, this new model is supposed to have a bigger caliber, as well. The NATO and Investigation name is "Black Eagle," as they can only speculate about the official name for it.

5

The Self-propelled Artillery of the USSR

In World War II, self-propelled guns had already proved to be effective. In addition, their production costs were lower than those of tanks. Through their limited area of aiming, though, they remained a support weapon. They were inferior to tanks for confrontation combat and pursuit into the depths of the enemy defenses.

The first rifle divisions of the postwar era had a mixed tank/self-propelled gun regiment, and each rifle regiment had a battery of SU-76 self-propelled guns. Through the exchange of old combat technique for new and the inclusion of tanks and self-propelled guns in the divisions and regiments, the weight of the artillery and grenade-launcher salvo rose to 3.5 tons by the end of 1946, in comparison with 1.589 tons in 1944. The number of shells fired per minute rose from 491,000 to 652,000.[20]

The artillery units equipped with self-propelled guns had been set up for the direct support of the tank and motorized rifle regiments. For a time, though, many self-propelled guns were used in the tank regiments as substitutes for the tanks that were lacking as a result of the increased numbers of tank units.

The fifties and sixties were typified chiefly, on the one hand, by more and more self-propelled guns being replaced by tanks, so that the mixed tank/self-propelled gun units became pure armored divisions, and on the other hand by more and more attention given to the question of whether armored artillery vehicles were necessary.

By the middle of the sixties, a series of various self-propelled artillery guns had been developed. None of these models, though, came to be introduced in significant numbers. The majority of the proposed prototypes were built on the chassis of World War II armored vehicles. But there were also attempts to use the hulls of the newly developed tanks for them.

By the start of the seventies, then, the Soviet military science finally equaled that of the international trend in introducing into the artillery units of the motorized rifle and armored divisions a relatively lightly armored tracked vehicle, but one armed with a large-caliber tank gun or howitzer. That this process took so much time resulted, among other things, from the hasty development and motorization of the artillery units; having so many tracked tractors, they were extremely mobile, even in rough terrain, and could regularly follow the attacking units. This resulted in not unjustified doubts about the practicality of artillery on armored vehicles. What with the conditions of a possible rocket and nuclear-weapon war, though, the advantages were paramount. Along with the protection of the crew from radioactivity, hand-weapon fire and splinters, it was possible to equip self-propelled guns with better observation and fire-control devices than towed guns. At the same time, the installation of automatic ammunition feed was possible, which increased the rate of fire and let the personnel be reduced. In addition, the self-propelled guns required a shorter time for being made ready to fire. Now as before, the disadvantages of higher manufacturing costs, more servicing, sensitivity of the complicated mechanisms, time-consuming crew training, and loss of the guns on account of minor drive-train damage remained.

To the present day, attempts at further development to increase firepower, improve mobility, increase the ammunition supply, reduce the dimensions of the guns and the basic vehicles, and increase their active radius have continued.

On the chassis of the T-54 medium tank, the design bureau in Kharkov developed the SU-122-54 medium self-propelled gun mount (1949 model). It was only produced for two years, starting in 1954.

20. Sakharov, M. W., Die Streitkräfte der UdSSR, Berlin 1968, p. 609.
21. Hell, W., SFL-Waffen zwischen Panzern und gezogenen Geschützen, Berlin 1983, p. 125ff.

Russian Tanks and Armored Vehicles 1946-to the Present

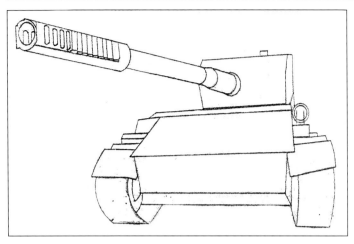

On the chassis of the IS heavy tank arose the ISU-152 heavy self-propelled gun mount (1945 model), developed as "Object 704."

Also designed in 1949 was the SU-152P medium self-propelled gun mount (1949 model), with a 152 mm gun, 5 crewmen, 28.7 ton weight, 7340 mm length, 3120 mm width, 2570 mm height, 400 HP Diesel engine, 55 kph top speed and 300 km range.

Medium Self-propelled Gun SU-122-54
Sdrednaya samochondo-artilleriyskaya ustanovka SU-122-54 (obr. 1949 g.)

Made in: USSR (tank factory in Kharkov)
Used in: USSR
Developed: 1949
Manufactured: 1954-1956
Crew: 5 men.
Fighting weight: 36.0 tons
Overall length: 9970 mm
Hull length: 6200 mm
Width: 3270 mm
Height: 2060 mm
Armament: 1 122 mm M-62-T cannon (35 rounds)
         1 14.5 mm KPWT heavy machine gun (300 rounds)
         1 14.5 mm KPWT AA machine gun (300 rounds)
Powerplant: 12-cylinder W-2-54 Diesel engine (520 HP/382 kW)
Top speed: 48 kph
Range: 400 km

Heavy Self-propelled Gun ISU-152
Tyashelaya samochodno-artilleriyskaya ustanovka ISU-152 (obr. 1945 g.) "Object 704"

Made in: USSR (tank factory in Chelyabinsk)
Used in: USSR
Developed: 1945
Manufactured: 1945
Crew: 5 men
Fighting weight: 47.3 tons
Overall length: 8530 mm
Hull length: 6800 mm
Width: 3150 mm
Height: 2240 mm
Armament: 1 152 mm ML-20SM howitzer (20 rounds)
         1 12.7 mm DSchK heavy machine gun (300 rounds)
         1 12.7 mm DSchK AA machine gun (300 rounds)
Powerplant: 12-cylinder W-2-IS Diesel engine (520 HP/382 kW)
Top speed: 40 kph
Range: 180 km

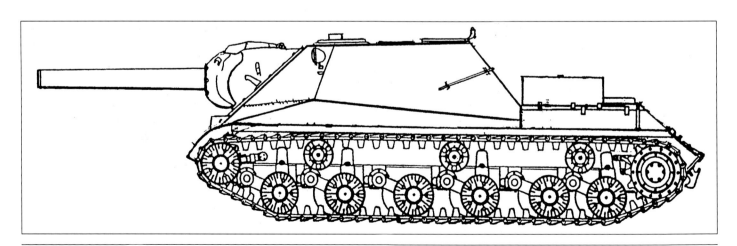

Chapter 5: The Self-propelled Artillery of the USSR

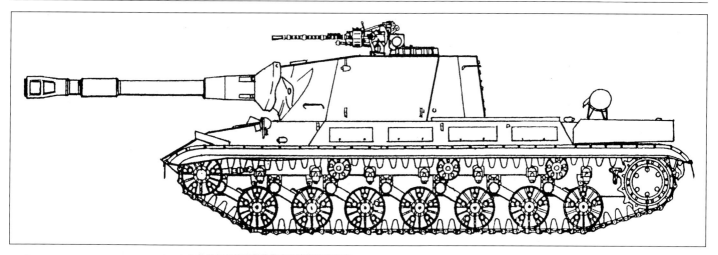

The tank works in Chelyabinsk designed the "Object 268" heavy self-propelled gun mount (1956 model) on the basis of the T-10 heavy tank, with a 4-man crew, 50 ton weight, 9100 mm length, 3270 mm width, 2675 mm height, 750 HP W-12-5 Diesel engine, 48 kph top speed and 350 km range.

Self-propelled Artillery Gun 2-S1 "Carnation"
Samochodno-artilleriyskaya ustanovka 2-SI "Gvosdika"
(obr. 1969 g.)

Made in: USSR/CIS/Russia, Bulgaria
Used in: USSR/CIS/Russia, Algeria, Angola, Bulgaria, CSSR, East Germany, Ethiopia, Hungary, Iraq, Libya, Poland, Syria, Yugoslavia, Finland
Developed: 1967-68
Manufactured: 1969 to date
Crew: 4 men
Fighting weight: 15.7 tons
Length: 7260 mm
Width: 2870 mm
Height: 2740 mm
Armament: 1 122 mm 2A31 howitzer (40 rounds)
Powerplant: 8-cylinder JaMZ 238W Diesel engine (315 HP/231 kW)
Top speed: 61.5 kph (road), 4.5 kph (water)
Range: 500 km

To fire large-caliber ammunition at long ranges, the Russian weapons industry built the 406.4 mm "Kondensator" howitzer cannon ("Object 271") on self-propelled mount 2A3 (1954 model), using a modified T-10 chassis. The shot range was 25 km. The muzzle velocity of the 470-kilogram shot was 716 m/s. The weight of the self-propelled mount was 64 tons.

Another mobile long-range gun was the 420 mm "Oka" ("Object 273") grenade launcher of self-propelled mount 2B1 (1957 model). Its range was between 25 and 45 kilometers, its fighting weight was stated at 55.3 tons.

Russian Tanks and Armored Vehicles 1946-to the Present

For the artillery battalions of the armored and motorized rifle divisions, the 2-S1 "Carnation" amphibious self-propelled artillery mount was built. The units equipped with BMP infantry combat vehicles were the first to receive this new artillery system. In the water it was propelled by its tracks.

A 2-S1 "Carnation" self-propelled artillery mount, used by a Czech artillery battalion.

Self-propelled Artillery Gun 2-S3 "Acacia", 2-S3M "Acacia"
Samochodno-artilleriyskaya ustanovka 2-S3 "Akaziya"
(obr. 1971 g.), 2-S3M "Akaziya" (obr. 1975 g.)

Made in: USSR/CIS/Russia
Used in: USSR/CIS/Russia, East Germany, Iraq, Libya, Syria
Developed: 1971
Manufactured: 1972 to date
Crew: 4 men
Fighting weight: 27.5 tons
Length: 7765 mm
Width: 3250 mm
Height: 3050 mm (with MG), 2615 mm (without MG)
Armament: 1 152.4 mm 2A33 howitzer (40 rounds), 2-S3M (46 rounds)
1 7.62 mm PKT AA machine gun (1500 rounds)
Powerplant: 12-cylinder W-59 multifuel Diesel engine (520 HP/382 kW), W-59U (580 HP/427 kW)
Top speed: 62 kph
Range: 500 km

Self-propelled Artillery Gun 2-S5 "Hyacinth"
Samochodno-artilleriyskaya ustanovka 2-S5 "Giazint"
(obr. 1979 s.)

Made in: USSR/CIS/Russia
Used in: USSR/CIS/Russia, Finland
Developed: 1974-75
Manufactured: 1979 to date
Crew: 5 men
Fighting weight: 28.0 tons
Overall length: 8330 mm
Hull length: 7765 mm
Width: 3250 mm
Height: 2760 mm
Armament: 152.4 mm 2A37 gun (30 rounds)
7.62 mm PKT machine gun (1,500 rounds)
Powerplant: 12-cylinder W-59 (multifuel) Diesel engine (520 HP/382 kW), W-59U (580 HP/427 kW)
Top speed: 62 kph
Range: 500 km

Artillery battalions, equipped with 2-S1 self-propelled guns, take firing positions under the cover of an artificial fogbank.

Chapter 5: The Self-propelled Artillery of the USSR

A 2-S1 "Carnation" (Russian "Gvosdika") self-propelled artillery mount of an East German artillery battalion.

In the artillery regiments of the armored and motorized rifle divisions of the soviet Army, two battalions used 2-S1 self-propelled guns and a third used the 2-S3 "Acacia" (1971 model) or 2-S3M (1975 model) "Acacia."

Self-propelled Artillery Gun 2-S7 "Peony", 2-S7M "Peony"
Samochodno-artilleriyskaya ustanovka 2-S7 "Pion" (obr. 1975 g.), 2-S7M "Pion"

Made in: USSR/CIS/Russia
Used in: USSR/CIS/Russia, CSSR
Developed: 1974-75
Produced: 1975 to date
Crew: 7 men
Fighting weight: 46.0 tons
Overall length: 13200 mm
Hull length: 7160 mm
Width: 3380 mm
Height: 2985 mm
Armament: 1 203.2 mm 2A44 gun (4 rounds)
           2-S7M (8 rounds)
Powerplant: 12-cylinder W-46-6 (multifuel) Diesel engine (792 HP/ 582 kW)
Top speed: 50 kph
Range: 500 km

The 2-S4 "Tulip" (Russian "Tyulpan") self-propelled artillery gun mount.

The 2-S3 self-propelled mount differed from the 2-S3M by having a more powerful engine and a larger ammunition supply. These 2-S3M "Acacia" (Russian "Akaziya") self-propelled guns are on the march.

69

The 2-S5 "Hyacinth" (Russian "Giatsint") self-propelled artillery mount was made for the artillery brigades of the Soviet/CIS/Russia army.

For use in the heavy artillery brigades, the 2-S4 self-propelled artillery gun mount (1973 model) "Tulip" was built on the chassis of the 2-S3 mount. It was armed with a 240 mm grenade launcher. The rate of fire was one shot per minute, the maximum range of the 130 kg grenade was 9.7 kilometers.

The 2-S7M "Peony" (Russian "Pion") self-propelled artillery mount, used by the 7th Artillery Division of the former Czechoslovakian Army.

In order to increase the shot range of the 152.4 mm grenades, the 2-S5 "Hyacinth" self-propelled artillery gun mount carried a gun that extended over the vehicle's entire length.

For use in the artillery divisions of the military districts, the 2-S7 (1975 model) and 2-S7M "Peony" self-propelled artillery mounts were armed with the 203.2 mm 2A44 cannon.

Chapter 5: The Self-propelled Artillery of the USSR

Between 1988 and 1990, the CIS forces began to be equipped with the new 2-S19 "Msta-S" self-propelled artillery mount, named after a Russian river.

With the 152 mm cannon, the 2-S19 attained a shot range of almost 36 kilometers using accelerated ammunition. Ordinary artillery shells could be fired to a distance of 24.7 kilometers.

Sometimes the units created their own self-propelled artillery mounts by mounting the guns on the beds of trucks or the tops of armored vehicles. This improvised artillery mount with an 82 mm grenade launcher uses the MT-LB multipurpose towing and transport vehicle in the mountains of Afghanistan.

As an antitank weapon for the motorized rifle divisions equipped with the BTR-60/70/80 armored transporters, the 2-S23 self-propelled mount was created, using the chassis of the BTR-80 and the turret of the 2-S9 "Anona" self-propelled airborne gun mount.

6

Anti-aircraft Tanks and Mobile Anti-aircraft Rockets

The Soviet military theorists wrote in the sixties about the significance of air-defense troops: "The troops of the air defense of the land combat forces are meant to defend the units, groups and large groups of the land combat forces and their backline services against the effects of enemy air action. They include anti-aircraft rockets, anti-aircraft artillery and radar troop units. In close collaboration with fighter-plane forces, they protect the motorized rifle, tank and artillery troop units reliably and constantly from enemy air attack and thus contribute to letting them enter into combat in a timely and orderly manner so they can fulfill their tasks."[22]

Although all army commands had already known in World War II that armored vehicles with anti-aircraft weapons had to be present with the combat troops, the equipping of the Red Army and Soviet Army with these systems was neglected for many years. In the fifties the army troops were still insufficiently equipped with flak artillery. To be sure, since 1944 all battle tanks had been fitted with an anti-aircraft machine gun, and firepower had been increased significantly through the introduction of new anti-aircraft guns, but mobile anti-aircraft weapons practically did not exist at all. This was a particular problem for the rifle divisions. "Through the motorizing of the troops and their equipping with tanks and self-propelled guns, the difference between the rifle and mechanized units vanished more and more. A weak link inside the organization of the rifle divisions remained, though: the insufficient numbers of flak artillery. . . ."[23]

To be able to make up for the lack of anti-aircraft weapons in the units of the land combat forces as well, at least in the short term, the BTR-40 and BTR-152 armored transporters, introduced in 1950, were fitted with twin 14.5 mm anti-aircraft machine guns and turned over to the troops as ZPTU-2 anti-aircraft guns on BTR-152. The designers at the Kharkov Tank Works received from the Defense Committee a contract to develop an anti-aircraft tank. Since no practical experience was at hand, as up to this time only a few ZSU-37 self-propelled anti-aircraft guns had been built in 1944, the development of a self-propelled armored anti-aircraft gun went on until 1955. After five years of development, the Soviet Army finally received the first AA tank, designated ZSU-57-2. In the same year it was suggested that the BTR-152 armored transporter be fitted with quad 14.5 mm AA machine guns, and the BTR-50P armored transporter be equipped with twin 14.5 mm AA guns as the ZTPU-2 self-propelled AA gun and with quad guns as the ZTPU-4. But series production never happened. Tests with these vehicles must have provided vital experience for the development of a self-propelled AA gun with high firepower. According to Soviet sources, production of the ZSU-57-2 is supposed to have continued until 1960. In the same year, production of the ZSU-23-4 self-propelled AA gun began. It became the self-propelled AA gun of all the USSR's allies, and is still used today in versions with numerous improvements. In the ZSU-23-4W version (1968 model), the turret front was modified in connection with a system of protection against nuclear weapons. The ZSU-23-4W1 (1971 model) differed from its forerunner only in that its command device was now mounted vertically. The ZSU-23-4M (1977 model) can be recognized by the additional box on the right side of the turret roof. It contains spare equipment for the radio measuring device.

Only in the mid-seventies was a successor suggested by the Russian armament industry; as of 1979 it finally, as the 2-S6 "Tunguska" self-propelled AA gun, reached the ground troops for anti-aircraft defensive use.

In the years between 1960 and 1980, the anti-aircraft units were supplied with numerous rocket systems. This led to a constant change in the equipment of the AA troops. In the early eighties, the first structures emerged, most of them being retained into the most recent past.

An anti-aircraft platoon was assigned to each tank and motorized rifle regiment. As a rule, it consisted of four ZSU-23-4 "Shilka" and four 9P31 (9P31M) launching vehicles (NATO code: SA-9/GASKIN) of the 9K31 "Strela-1" anti-aircraft rocket complex. Both systems were meant to come directly into action to defend the combat troops and during marches. The "Shilka" fought low-flying targets up to 2,500 meters above ground, and the "Strela-1" took on those at ranges from 900 to 4,200 meters flying by and from 30 to 3,500 meters up while flying over. With the "Tunguska" complex, the two weapon systems were united in one mobile station for the first time. The effective range of the two 30 mm 2A38 or 2A38 M

---

22. Defense Ministry of the USSR, Militart Dictionary, Moscow, no date, p. 976.
23. Sacharov, M. W., Die Streitkräfte der UdSSR, Berlin, 1968, p. 610.

Chapter 6: Anti-aircraft Tanks and Mobile Anti-aircraft Rockets

cannons was 4,000 meters, and that of the 9M311 anti-aircraft rockets (NATO code: SA-19/GRISON) was 2,500 to 8,000 meters.

As the eighties began, the "Strela-1" began to be replaced by the 9K35 or 9K37 "Strela-10" AA rocket complex, with the 9A34 (9A34M) or 9A35 (9A35M) launching vehicle (NATO code: SA-13/GOPHER). Here the 9A34, 9A34M, 9A35 and 9A35M launching vehicles can also be equipped to fire the AA rockets of the 9K31/9K31M complex. The last launching vehicles were equipped for the 9K37 AA rocket complex. This can fire the 9M31, 9M31M, 9M37 and 9M37M rockets as well as the newest development, the 9M333. The fact that the anti-aircraft rockets are interchangeable has led to astounding designations in Western literature; we need only note the NATO code (SA-9)/SA-13/(GASKIN)/GOPHER.

Every three of these regiments was joined at the division level by an anti-aircraft regiment of five batteries. Each of these batteries had one 2K12 "Kub" (2K12M "Kub-M," 2K12M2 "Kub-M2" and 2K12M3 "Kub-M3") AA rocket complex, consisting of four 2P25 (2P25M and 2P25M2) launching vehicles (NATO code: SA-6/GAINFUL), one 1S91 reconnaissance and guiding station (later 1S91M and 1S91M2; NATO code: STRAIGHT FLUSH), and the appropriate technical defenses. Every launching vehicle can fire three 3M9 (3M9M, 3M9M2, 3M9M3) AA rockets. The effective range for attacking aircraft flying past is between 4,000 and 35,000 meters. The maximum height of overflying air targets is some 11,000 meters.

During a parade in Moscow on November 7, 1975, the 9A33 launching and guiding station (NATO code: SA-81GECKO) was first shown to the public. As part of the 9K33 "Osa" anti-aircraft rocket complex, it has been used since its introduction in AA rocket batteries of the motorized rifle and tank divisions. Every battery, as a rule, has two platoons, each with two launching and guiding stations, as well as one 9T217 transport and freight vehicle. As a mobile AA rocket complex, it serves to protect units on the march, during attack and in defense. The system completes the defenses of the "Kub" complex and attacks low-flying targets from 25 to 6,000 meters up. The target distance can be as much as 10,000 meters. The complex has been improved several times to date. In 1980 the Soviet Army presented the improved 9A33B launching and guiding station of the "Osa-A" complex, with 9M33M AA rockets in containers (NATO code: SA-81GECKO). At the same time, the 9A33BM2 launching and guiding station of the "Osa-AK" complex began to be introduced. In the mid-eighties there followed the 9A33BM launching and guiding station of the 9K33M2 "Osa-AK" complex, and at the end of the eighties the 9A33BM3 launching and guiding station of the 9K33M3 "Osa-AKM" complex. All vehicles equipped with containers have the NATO code of SA-8b/GECKO and cannot be told apart externally. The improvements concerned chiefly the targeting means, and target-following radar, the guiding and control systems, the friend-foe recognition devices, and the rockets themselves. At the end of the eighties, an antenna was added to the radar scope for the new friend-foe recognition device. Until then the 1S51 device (improved to 1S51M3-2) was utilized.

In the AA brigades of the armored, rifle and shock armies, the 2K11 "Krug" AA rocket complex was put into use. Every brigade had three batteries with three 2P24 launching vehicles, one 1S32 rocket guiding station (NATO code: PAT HAND), and one 1S12 reconnaissance and targeting station (NATO code: LONG TRACK). This anti-aircraft defense system was used in the backland areas of the armies and was meant to destroy enemy aircraft at altitudes up to 27,000 meters. The targeting range for craft flying past is stated as 70,000 meters.

As an answer to the air- and sea-supported march rocket introduced by NATO in the seventies, the designers of the Russian armament industry developed a launching vehicle on the MAZ-7310 truck (NATO code: SA-10/GRUMBLE) for the S-300 AA rocket complex, the 9A82 (NATO code: SA-12a/GLADIATOR) and 9A83 launching vehicles (NATO code: SA-12b/GIANT), and the S-300W AA rocket complex (NATO code: SA-11 GADFLY). These complexes included, in addition to the launching vehicles, a series of radar trucks for seeking, finding, pursuing and illuminating targets and guiding rockets. These anti-aircraft defense systems make it possible to attack several aircraft simultaneously or to fire two AA rockets at one target.

A rather late beginning was made in mounting anti-aircraft weapons on armored vehicles. To make up quickly for this lack, numerous armored transporters were fitted with heavy machine guns for anti-aircraft defense in the fifties.

The ZPTU-2 self-propelled AA M.G. (on BTR-40).

The ZPTU-2 self-propelled AA M.G. (on BTR-152), used by a unit of the former East German Army, which designated the vehicle as the SPW-152A armored troop carrier.

The ZPTU-2 (Russian: Zenitnaya potivnaya tankovaya ustanovka ZPTU-2 na BTR-152 obr. 1950 g.) self-propelled AA M.G. (1955 model). Its two KPWT 14.5 mm heavy machine guns were removed.

Self-propelled Anti-aircraft Gun ZSU-57-2
Zenitnaya samochodno-artilleriyskaya ustanovka ZSU-57-2 (obr. 1955 g.)

Made in: USSR (tank factory in Kharkov)
Used in: USSR, Algeria, Angola, Bulgaria, Cambodia, China, CSSR, Cuba, East Germany, Egypt, Ethiopia, Finland, Hungary, Iran, Iraq, Korea, Mongolia, Mozambique, North Korea, Poland, Romania, South Korea, Syria, Vietnam, Yugoslavia
Developed: 1955
Manufactured: 1955-1960
Crew: 5 men
Fighting weight: 28.0 tons.
Overall length: 8480 mm
Hull length: 6220 mm
Width: 3720 mm
Height: 2750 mm
Armament: 2 57 mm S-68 L/73 AA guns (316 rounds)
Powerplant: 12-cylinder W-2-54 Diesel engine (520 HP/382 kW)
Top speed: 48 kph
Range: 420 km

Self-propelled Anti-aircraft Gun ZSU-23-4 "Shilka", ZSU-23-4W "Shilka", ZSU-23-4W1 "Shilka", ZSU-23-4M "Shilka"
Zenitnaya samochodno-artilleriyskaya ustanovka ZSU-23-4 "Shilka" (obr. 1960 g.), ZSU-23-4W "Shilka" (obr. 1968 g.), ZSU-23-4W1 "Shilka" (obr. 1971 g.), ZSU-23-4M "Shilka" (obr. 1977 g.)

Made in: USSR (tank factory in Volgograd)
Used in: USSR/CIS/Russia, Afghanistan, Algeria, Angola, Bulgaria, Cambodia, CSSR, Cuba, East Germany, Egypt, Ethiopia, Hungary, India, Iran, Iraq, Jordan, Libya, Mozambiaue, Nigeria, North Korea, North Yemen (YAR), Peru, Poland, Somalia, South Korea, Syria, Vietnam, Yugoslavia
Developed: 1960
Manufactured: 1960-1985/86
Crew: 4 men
Fighting weight: 19.0 tons (ZSU-23-4), 19.2 tons (ZSU-23-4W/W1), 21.0 tons (ZSU-23-4M)
Length: 6536 mm
Width: 3125 mm
Height: 3572 mm (ready to fire), 2644 mm (ready to march)
Armament: 4 23 mm AZP-23 (AZP-23M) AA guns (2000 rounds)
Powerplant: 6-cylinder W-6R (W-6R-1, W-6M, W-6M-1) Diesel engine (240 HP/177 kW, 280 HP/206 kW)
Top speed: 50 kph
Range: 325 km

A ZSU-57-2 self-propelled AA gun of a former East German Army unit, during firing practice in August 1966.

Chapter 6: Anti-aircraft Tanks and Mobile Anti-aircraft Rockets

The ZSU-57-2 self-propelled AA gun (1955 model), used by its crew in training in the normal times for mounting and dismounting.

In the improved ZSU-23-4W self-propelled AA gun mount (1968 model), a modified turret was used in connection with the installation of a nuclear weapon protection system. The intake shafts with filtering ventilation apparatus for the air intake were situated on both sides of the turret front beside the tank gun. These Polish Army vehicles are crossing a pontoon bridge.

Self-propelled Anti-aircraft gun 2-S6, 2-S6M "Tunguska"
Zenitnaya samochodno-artiulleriyskaya ustanovka 2-S6, 2-S6M "Tunguska"

Made in: USSR/CIS/Russia
Used in: USSR/CIS/Russia, Indien
Developed: 1974-75
Manufactured: 1978-79 to date
Crew: 4 men
Fighting weight: 34.0 tons
Length: 7930 mm
Width: 3240 mm
Height: 4020 mm
Armament: 2 30 mm 2A38 (2A38M) AA guns (1900 rounds)
                 8 9M311 AA rockets
Powerplant: 12-cylinder W-59 (multifuel) Diesel engine (520 HP/ 382 kW), later W-84A (842 HP/619 kW)
Top speed: 60 kph (2-S6), 65 kph (2-S6M)
Range: 500 km

In the ZSU-23-4W1 self-propelled AA gun mount (1971 model), a new command device was installed to control the weapon system. It was only installed vertically. The previous model had been installed horizontally.

The ZSU-57-2 self-propelled AA gun mount was first shown to the public in a Moscow parade in November 1957.

In the Moscow parade on November 7, 1965, the ZSU-23-4 (1960 model) "Shilka" self-propelled AA gun (named after a Russian river) was first shown to the public.

On the ZSU-23-4M self-propelled AA gun (1977 model), an additional rack for the radar device was installed on the roof of the turret.

The ZSU-23-4M self-propelled AA gun (1977 model) with the antenna (covered) in marching position.

Launching Vehicle 9P31, 9P31M
Raketnaya samochodno-puskowaya ustanowka 9P31, 9P31M

Made in: USSR
Used in: USSR, Algeria, Angola, Benin, Bulgaria, CSSR, East Germany, Egypt, Guinea, Guinea-Bissau, Hungary, India, Iraq, Israel, Madagascar, Mali, Mauritania, Mozambique, Nicaragua, Poland, South Africa, South Yemen, Syria, Tanzania, Vietnam, Yugoslavia
Developed: 1964-65
Manufactured: 1966-1975/76
Crew: 2 men
Fighting weight: 7200 kg
Length: 5750 mm
Width: 2350 mm
Height: 3990 mm (ready to fire), 2550 (ready to march)
Armament: 4 9M31 or 9M31M AA rockets
Powerplant: 8-cylinder GAZ-41 gasoline engine (140 HP/103 kW)
Top speed: 100 kph (road), 10 kph (water)
Range: 750 km

Launching Vehicle 9A34, 9A34M, 9A35, 9A35M
Raketnaya samochodno-puskowaya ustanowka 9A34, 9A34M, 9A35, 9A35M

Made in: USSR/CIS
Used in: USSR/CIS/Russia, Algeria, East Germany, Cuba, Iraq, CSSR, Libya, Poland, Syriaa
Developed: 1974-75
Manufactured: 1975-88/89
Crew: 3 men
Fighting weight: 12,130 kg
Length: 6454 mm
Width: 2850 mm
Height: 3860 mm (ready to fire), 2220 mm (ready to march)
Armament: 4 9M31, 9M31M, 9M37, 9M37M of 9M333
Powerplant: 8-cylinder JaMZ-238W Diesel engine (240 HP/177 kW)
Top speed: 51.5 kph (road), 6 kph (water)
Range: 500 km

The 2-S6M "Tunguska" self-propelled AA gun mount (named after a Russian river) had a four-man crew, was armed with two 30 mm 2A38M AA guns (and carried 1,900 rounds) and eight 9M311 AA rockets (NATO code: SA-19/GRISON). The rate of fire of the two 30 mm guns on the 2-S6M "Tunguska" self-propelled mount was stated at a combined shots 4,800 per minute. The practical shot range for approaching and departing targets was given as 4,000 meters, with their height at up to 2,000 meters and the distance to a target flying past also up to 2,000 meters.

Chapter 6: Anti-aircraft Tanks and Mobile Anti-aircraft Rockets

The 9P31 launcher for the 9K31 "Arrow-1" (Russian "Strela-1") AA rocket complex carried four 9M31 or 9M31M rockets (NATO code: SA-9/GASKIN).

The 9P31M lanucher for the 9K31 "Arrow-1" AA rocket complex carried four 9M31M rockets (NATO code: SA-13/GASKIN).

Launching Vehicle 9A33, 9A33B, 9A33BM, 9A33BM2, 9ABM3
Raketnaya samochodno-puskowaya ustanovka 9A33, 9A33B, 9A33BM, 9A33BM2, 9A33BM3

Made in: USSR/CIS/Russia
Used in: USSR/CIS/Russia, Algeria, Angola, East Germany, Guinea-Bissau, Hungary, India, Iraq, Jordan, Kuwait, Libya, Poland, Syria, Yugoslavia
Developed: 1975-76
Manufactured: 1977-78 to date
Crew: 3 men
Fighting weight: 18,000 kg
Length: 9270 mm
Width: 2830 mm
Height: 4853 mm (ready to fire), 3998 mm (ready to march)
Armament: 4 9M33 AA rockets (only on 9A33 launcher)
        6 9M33M or 9M33M2 AA rockets (on 9P35 launching apparatus)
Powerplant: 6-cylinder UTD-20 Diesel engine (295 HP/2117 kW)
Top speed: 60 kph (road), 8 kph (water)
Range: 800 km

Launching Vehicle 2P25, 2P25M, 2P25M2
Raketnaya samochodno-puskowaya ustanovka 2P25, 2P25M, 2P25M2

Made in: USSR
Used in: USSR/CIS/Russia, Algeria, Angola, Bulgaria, CSSR, Cuba, East Germany, Egypt, Ethiopia, Guinea, Hungary, India, Iraq, Libya, Mozambique, North Yemen, Poland, Romania, Somalia, Syria, Tanzania, Vietnam, Yugoslavia
Developed: 1959 (Toropov Design Bureau, Tank Works 134, Omsk)
Manufactured: 1960-61 to 1984-85
Crew: 3 men
Fighting weight: 20,600 kg
Overall length: 7389 mm
Hull length: 6810 mm
Width: 3040 mm
Height: 2818 mm
Armament: 3 3M9, 3M9M, 3M9M2, 3M9M3 AA rockets
Powerplant: 6-cylinder W-6R, W-6R-1, W-6M, W-6M-1 Diesel engine (240 HP/177 kW), 280 HP/206 kW)
Top speed: 50 kph
Range: 425 km

The 9A35M launcher for the 9K35M "Arrow-10" (Russian "Strela-10") AA rocket complex carried four 9M37M rockets (NATO code: SA-13/GOPHER).

77

The gunner's seat in the 9A35M launcher vehicle.

A 2P25 launcher of the 2K12 "Cube" (Russian "Kub") AA rocket complex, with three 3M9 rockets (NATO code: SA-6/GAINFUL).

Launching Vehicle 2P24
Raketnaya samochodno-puskovaya ustanovka 2P24

Made in: USSR
Used in: USSR/CIS/Russia, Bulgaria, CSSR, East Germany, Hungary
Developed: 1957-58
Manufactured: 1960-61 to 1979-80
Crew: 3 men
Fighting weight: 28,200 kg
Overall length: 9460 mm
Hull length: 7100 mm
Width: 3250 mm
Height: 4472 mm
Armament: 2 3M8, 3M8M, 3M8M2 AA rockets
Powerplant: 12-cylinder W-105, W-105W (multifuel) Diesel engine (400 HP/294 kW)
Top speed: 63.4 kph
Range: 360 km

The 9A33 launching and guiding station of the 9K33 "Wasp" (Russian "Osa") AA rocket complex, with four 9M33 rockets (NATO code: SA-8/GECKO).

Launchers on parade in Moscow.

A 2P25M launcher of an AA rocket unit of the former East German Army, seen from the rear.

Chapter 6: Anti-aircraft Tanks and Mobile Anti-aircraft Rockets

The 9A33BM2 launching and guiding station of the 9K33M2 "Wasp-AK" (Russian "Osa-AK") AA rocket complex, for six 9M33M rockets in containers, only two containers set up (NATO code: SA-8b/GECKO).

The launching apparatus, here with a 9M33M2 GWR rocket in a container, the screens of the target-following radar (center), and the fire-control radar (right and left).

AA rocket comples "Thor."

The 9Sch38-2 TV screen (for up to 25,000 meters) and rotating radar device with antenna (turning speed 33 rpm).

The 9A33BM3 launching and guiding station with added antenna for the "Password" friend-foe recognition device.

Shown for the first time in Moscow on May Day 1964: the 9P24 launcher for the 2K11 "Circle" (Russian "Krug") AA rocket complex, with two 3M8 rockets (NATO code: SA-4/GANEF).

The 2P24M launcher of the 2K11M "Circle-M" (Russian "Krug-M") AA rocket complex, with two 3M8M rockets (NATO code: SA-4b/GANEF).

The launcher for the S-300 AA rocket complex (NATO code: SA-10b/GRUMBLE).

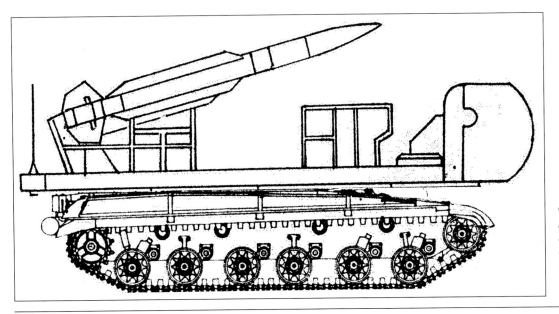

The lanucher for the AA rocket complex developed at the start of the nineties, with four rockets (NATO code: SA-11/GADFLY); four-man crew, fighting weight circa 16 tons (the sketch shows the author's conception).

Chapter 6: Anti-aircraft Tanks and Mobile Anti-aircraft Rockets

The 9A82 launcher for the S-300W AA rocket complex, with two rockets in containers (NATO code: SA-12a/GLADIATOR).

The 9A83 launcher for the S-300W AA rocket complex, with four rockets in containers (NATO code: SA-12b/GIANT).

7

Mobile Multipurpose Launchers - in Russian, "Battle Machines"

At the end of World War II, there were 519 launcher units in the Russian Army; they were also known as guard launcher units. In all, there were 105 artillery regiments, 40 independent brigades and seven divisions equipped with these launchers. The share of launchers in the reserve artillery of the high command of the Red Army amounted to 13 per cent. Scarcely any other weapon, other than the T-34 tank, was as celebrated as the "victor's weapon" as these "Katyuschas." Among the Red Army's opponents, they were known as "Stalin Organs" and were much feared.

Even though the scene has become quieter in the present where this weapon system is concerned, they are now as then an essential part of the self-propelled artillery. "Missile launchers are combat vehicles of the rocket-troop service arm and artillery of the ground combat forces; they carry multiple launching apparatus for unguided reactive missiles, are mobile as a rule, and thus can be regarded as self-propelled mounts. Since the missile launchers are made so that they can fire a large number of missiles in a very short time, they are also known as salvo firing systems."[24]

As the forties turned to the fifties, the Soviet Army continued to use the BM-13 and BM-31 missile launchers of the last war. But the trucks that had come from the USA within the parameters of the lend-lease agreement had to be returned. So as not to run out of missile launchers, they were mounted on ZIS-5 and ZIL-151 trucks. There was also a series of improved missiles, launching and targeting devices, as well as launching vehicles. The missiles were improved in terms of range, less scattering, and a greater effect on the target.

Typical of the postwar development of the missile launcher was the giving up of launching rockets on launching rails. The missile-launcher rockets were now launched from tubes or tubelike devices.

For the turboreactive M-14 OF 140 mm splinter explosive shell (original weight 39.6 kg, range 9,500 meters), the BM-14 launcher, with sixteen launching tubes, was mounted on the chassis of the Zil-151 (6x6) medium off-road truck. The time required to change from marching to readiness to fire was only 1.5 to 2 minutes. The reloading time was to be only two to three minutes. The BM-14-16 was used in the independent missile-launcher units of the motorized rifle divisions in the fifties.

At the beginning of the sixties the BM-14 was modernized. Essentially, only shorter tubes were used. This was made possible through the use of more powerful launching charges. With the refinement of mount design, the gross weight of the launching apparatus decreased to 2,800 kilograms, despite the added launching tubes. This allowed the BM-14-17 to be mounted on the chassis of the GAZ-63 (4x4) light off-road truck.

In 1954, the Soviet Army introduced the BM-24 launcher at the October parade in Moscow. It fired turboreactive rockets with a caliber of 240 mm. The maximum range was some 10,000 meters. The launching weight of a rocket was 110 kilograms.

Three years later, the BM-24T rolled across Red Square in Moscow. Built on the chassis of the AT-S artillery tractor, it had been developed especially for the rocket-launcher units which were to see action in the armored divisions.

With the introduction of the BMD-20 missile launcher on the ZIL-157 (6x6) medium off-road truck, the Soviet Army obtained a medium-range missile launcher. The maximum range of the 200 mm rockets was 19,000 meters. In order to keep the dispersion relatively meager at this great distance, the four launching apparatus were made in spiral form. The shot rotation was also supported by a ring of jets set around the main jet and inclined at six degrees from the long axis of the rocket.

To fire on surface targets at ranges up to 55,900 meters, the BM-28 missile launcher was installed on the KrAZ-214 (6x6) heavy truck.

To this day the BM-21 missile launcher on the Ural 375 or Ural 4320 medium off-road (6x6) truck is among the standard weapons of the missile-launcher units of the motorized rifle and armored divisions of the Soviet Union, its successors and allies. One unit, as a rule, consists of three batteries with six launchers each. The BM-21 launchers were first shown to the public at the moscow parade on November 7, 1964. In all, forty tubes with a caliber of 122.4 mm were combined into one package. The maximum range is stated as 20,000 meters. A complete salvo of forty rockets can be fired in twenty seconds. M-21 OF reactive splinter explosive missiles, with

---

24. Kiesshauer, W., Geschosswerfer—eine maechtige Artilleriewaffe, Berlin 1984, p. 1

Chapter 7: Mobile Multipurpose Launchers - in Russian, "Battle Machines"

a length of 2870 mm and weight of 66 kilograms are used. The weight of the warhead is 20.45 kilos, of which 6.4 kilos are the explosive charge.

The BM-21W missile launcher on the GAZ-66 light off-road (4x4) truck has been in production since the early seventies. From twelve tubes it fires the same rockets as the BM-21, and was introduced primarily for the airborne troops.

Since the mid-seventies, the BM-22 "Uragan" (Hurricane) missile launcher on the ZIL-135 heavy off-road (8x8) truck (Launching Vehicle 9P140) has been used by the launcher units of the artillery divisions of the USSR/CIS/Russia, its successors and allies. The caliber of the sixteen launching tubes is 220 mm. A rocket is some 4600 mm long and weighs 360 kilograms. The maximum range is about 35,000 meters.

During the 1990 weapons fair in Kuala Lumpur, Malaysia, the Russian armament industry demonstrated the 9A52 "Smerch" (Whirlwind) missile launcher. Since 1987 the introduction of this weapons system has become internationally known. The twelve-tube launcher with a caliber of 300 mm is mounted on the chassis of the MAZ-7310 heavy off-road (8x8) truck. With a range between 20,000 and 75,000 meters, both conventional rockets and all other types of ammunition can be fired.

Missile Launcher BM-24T
Boewaya maschina BM-24T

Made in: USSR
Used in: USSR
Developed: 1954-55
Manufactured: late fifties
Crew: 7 men
Fighting weight: 15.0 tons
Length: 5870 mm
Width: 2570 mm
Height: 2700 mm
Armament: 12 240 mm BM-24 launching tubes (12 reactive missiles)
Powerplant: 12-cylinder W-2-T Diesel engine (275 HP/202 kW)
Top speed: 35 kph
Range: 380 km

The BM-13 projectile launcher (on a ZIL-151 truck).

Until about 1980, the BM-13 projectile launcher was among the equipment of the Soviet Army units stationed east of the Ural Mountains.

BM-14-16 projectile launchers (on ZIL-151 trucks) of a unit of the group of Soviet forces in Germany.

Loading the 39.6-kilogram rockets into the BM-14-16 launcher.

Salvo fire from BM-14-17 launchers.

The BM-24 launcher (on a ZIL-151 truck), seen during a parade in Moscow.

BM-24 launchers (on ZIL-157 trucks) of an East German Army unit on the march.

The BM-14-17 projectile launcher (on a GAZ-63A truck).

Chapter 7: Mobile Multipurpose Launchers - in Russian, "Battle Machines"

BM-24T launcher (on AT-S medium artillery tractor).

The BMD-20 launcher (on a ZIL-151 truck).

The BM-21W launcher (on a GAZ-66 truck).

The BM-25 launcher (on a KrAZ-214 truck).

The BM-24T launchers were intended mainly for use in the Soviet Army's armored divisions.

The standard vehicle of the Soviet launcher units—the BM-21 (on a Ural 375 truck).

85

According to the shot values, the gunner sets the range on the K-1 Kollimator.

The BM-21 launcher consists of forty tubes of 122.4 mm caliber.

The BM-21 "Hail" launcher (Russian "Grad") (on a Ural 4320 truck).

The 9P140 launching vehicle with BM-22 "Hurricane" launcher (Russian "Uragan") (on a ZIL-135 truck).

The 9A52 launching vehicle with "Windstorm" (Russian "Smerch") launcher (on an MAZ-7310 truck).

8

The Mobile Rocket Carriers of the Soviet Army

On July 15, 1946, the first rocket troop unit of the Soviet forces was developed out of a guard launcher regiment. Tests in 1947 with the first rockets developed after the war showed the possibilities of their military uses. Their great speed and extreme altitude were to guarantee invulnerability. The rockets were regarded as the most highly developed and best-suited carriers for atomic ammunition. Increasing their range and accuracy were regarded as the main goals of their development. Within a very short time, rocket types of varying sizes were developed by the USSR's armament industry. After the first intercontinental ballistic multi-stage rockets had been tested successfully in August 1957 and the first artificial earth sattelite had been put into orbit in the same month, the rocket troops were divided into three parts. Since then, the ground combat troops have included the units equipped with tactical and operative-tactical rockets. While the tactical rockets have a range of up to 120 kilometers, the operative-tactical kinds can reach about 300 kilometers. In military technical language they are thus designated R-65 (NATO code: FROG) and R-300 (NATO code: SCUD).

The strategic rocket units are directly subordinate to the high command of the Soviet armed forces. Until 1963-64, the targeting precision of the strategic rockets was improved so much that they scarcely strayed from targets 12,000 to 13,000 kilometers away. The practically unlimited range of the strategic rockets and the possibility of extremely varied maneuvers during flight guaranteed that any target on earth could be reached from Russian territory. The state and party leadership of the USSR regarded the rocket troops as *the* means of inspiring fear among the Western forces. "The decisive means of war are presently the strategic nuclear weapons and their wide-ranging carriers. The equipping of the Soviet forces with mighty, long-range strategic weapons allows direct attacks on the enemy's strategic nuclear weapons, his economic potential and his political and military command system. Any state, especially one with a small, heavily populated territory, can be knocked out and even wiped out in a short time, without ground combat forces having to enter its territory."[25] Within a few years, numerous new units were established. At the parade in Moscow on November 7, 1965, smaller intercontinental rockets with solid fuel on self-propelled launching ramps were displayed for the first time. Such rocket systems were not limited to permanent launching positions and thus could not be observed in advance by air and cosmic reconnaissance, but were ready for action within a few hours.

Launching Vehicle 2P16
Raketnaya samochodno-puskowaya ustanovka 2P16

Made in: USSR
Used in: USSR, CSSR, East Germany, Poland
Developed: 1955/56
Manufactured: 1957-1965
Crew: 3 men
Fighting weight: 17.2 tons
Overall length: 9250 mm
Hull length: 7250 mm
Width: 3140 mm
Height: 2990 mm
Armament: 1 9M21 "Luna" tactical rocket (R-65)
Powerplant: 6-cylinder W-6 Diesel engine (240 HP/177 kW)
Top speed: 45 kph
Range: 250 km

The 2P16 launcher with "Moon" tactical rocket (NATO code: FROG).

25. Sokolovski, W. D., The Military Strategy of the USSR, Moscow 1975.

Launching of an R-65 tactical rocket.

The 2P16 launcher with 9M21 "Moon-M" tactical rocket (NATO code: FROG-4).

Launching Vehicle 9P117
Raketnaya samochodno-puskovaya ustanovka 9P117

Made in: USSR, China
Used in: USSR, Bulgaria, CSSR, Cuba, East Germany, Ethiopia, Hungary, Iraq, Mongolia, Poland, Romania, Syria, Yugoslavia, North Korea, Iran, Lybia, Afganistan, Algeria, Yemen, Vietnam
Developed: 1961-62
Manufactured: 1964 to 1984-85
Crew: 4 men
Fighting weight: 37.4 tons
Length: 12,360 mm
Width: 3020 mm
Height: 3330 mm
Armament: 1 operative-tactical R-17 "Elbus" with rocket (R-14)
Powerplant: 12-cylinder D12A 525-A Diesel engine (525 HP/386 kW)
Top speed: 70 kph
Range: 500 km

Launching Vehicle for Medium-range Rockets RSD-10
Raketnaya samochodno-puskovaya ustanovka glya raketa sredney dalnosti RSD-10

Manufacturer: USSR (Votkinsk machine shop, Udmurtish ASSR)
Used in: USSR
Developed: ca. 1975
Manufactured: to 1988
Crew: 2 men
Fighting weight: 82.95 tons (launching vehicle 40.25 tons, rocket with container 42.7 tons)
Length: 16,810 mm (without rocket and container)
Armament: 1 RSD-10 medium-range rocket (R-5500)
Powerplant: MAZ Diesel engine
Top speed: 50 kph
Range: -

The 9M21 "Luna-M" tactical rocket at rest.

The 9M21 "Luna-M" tactical rocket in launching position.

Chapter 8: The Mobile Rocket Carriers of the Soviet Army

The 9P113 launcher for the 9K52 tactical rocket complex.

The 9P113 launcher fired the 9M21 "Moon-M" tactical rocket (NATO code: FROG-7).

The 9P129 launcher for the 9K79 "Little Dot" (Russian "Toshka") rocket complex.

The 8U218 launcher with 3R-1 "Owl" operative-tactical rocket. (Russian code "Filin", NATO code: FROG-1)

The 9P129 launcher with 9M79F, 9M79K or 9M79B tactical rocket (NATO code: SS-21/SCARAB).

The 2P19 launcher of the 8K12 operative-tactical rocket complex (NATO code: SS-1b/SCUD-A).

The 8U218 launcher of the 8K11 operative-tactical rocket complex (NATO code: SS-1/SCUD).

The 2P19 launcher of the 8K11 operative-tactical rocket complex, with 8K11 rocket (NATO code: SS-1/SCUD-A).

The 9P117 launcher could launch the 8K14 (R-300) rocket.

The 9P117 launcher of the 8K14 "Elbrus" operative-tactical rocket complex (NATO code: SS-1c/SCUD-B).

Chapter 8: The Mobile Rocket Carriers of the Soviet Army

The 9P117M launcher of the 9K72 operative-tactical rocket complex.

The 9P117M launcher with OTR-22 operative-tactical rockets (NATO code: SS-12/SCALEBOARD) in marching position.

The range of the strategic rockets (NATO code: SS-X-14/SCAMP) was 4,000 to 4,500 kilometers.

The 9P71 launcher of the 9K714 "Oka" (named after a Russian river) operative-tactical rocket complex.

The 9P71 launcher fired 9M714 operative-tactical rockets (NATO code: SS-23/SPIDER).

The launcher of the 8K96 strategic rocket complex in the May Day 1965 parade in Moscow.

The launcher of the 8K99 strategic rocket complex in Moscow on May Day 1965.

The strategic rocket (NATO code: SS-X-15/SCROOGE) had a range of up to 6,000 kilometers.

The launcher for long-range strategic rockets (NATO code: SS-25/SICKLE).

The RSD-10 medium-range rocket (Russian: "Raketa sredney dalnosti") could hit strategic targets at ranges up to 5,500 kilometers.

The launcher of the RSD-10 strategic rocket complex. (NATO code: SS-201SABER)

Chapter 8: The Mobile Rocket Carriers of the Soviet Army

The launcher for the coast guard rocket complex (NATO code: SSC-1b/SEPAL).

The launcher for coast guard rockets, on the ZIL-137 (8x8) truck chassis was first shown in 1965.

The launcher for the 'Harpoon' (Russian: "Garpun") coast guard rocket complex in Yugoslavia. The allowable gross weight was 30.5 tons, the length 13,950 mm, width 3150 mm, and height 4100 mm.

The "Waldemar Verner" Coast Guard Rocket Regiment 18 in Schwarzenpfrst, serving with the East German Navy, had ten launchers of the "Harpoon" coast guard rocket complex, designated "Rubesch Ä," or ZC-50, or ET-161.

From the two containers, the P-21 coastal rockets with radar target-seeking heads, and P-22, with infra-red target-seeking heads (NATO code: SSC-3/BROM or SSC-4/COAST), could be fired.

9

Tank Destroyers - The Tank-destroyer Guided Missiles on Motorized Launches

Until the introduction of multilayered armor, the hollow-charge shells were regarded as *the* armor-piercing ammunition of the 20th century. Since their penetrating power was independent of the shot range, the only problem seemed to be that of precise targeting. "In the mid-fifties, the antitank guided missiles were invented. They initiated a stormy development of highly effective weapons for use against armored targets. The ground combat forces were now equipped with antitank guided-missile complexes that could destroy any tank at a distance of two and more kilometers with a single shot."[26]

In the thirties there were already conceptions of very small rockets in the offices of rocket researchers. But only after the groundwork for the development of the German X-7 "Little Red Riding Hood" antitank rocket fell into Russian hands at the end of World War II were the experiments continued.

But until May 1, 1962, when the Tamaner motorized rifle division drove through Red Square in Moscow with the 2P26 and 2P27 launching vehicles for four and three electrically controlled 3M6 "Shmel" (Bumble-bee) antitank rockets, there was no knowledge of their stage of development. Since the end of the fifties, though, the launching vehicles for antitank guided missiles have belonged to the tank-destroyer companies of the motorized rifle divisions.

Two years later, the improved 3M11 "Falanga" (closed front) antitank rocket was displayed, mounted on the BRDM-1 reconnaissance vehicle. On May 9, 1965, for the twentieth anniversary of the victory, the adapted 9P110 launching vehicle for six 9M14 "Malyutka" (Tiny One) antitank rockets was displayed in Moscow. The new carrier vehicle no longer had its rear cover folding to the side, but raised along with the launching apparatus for the antitank rocket. Another new feature of this system was its semi-automatic launching. The missiles were electrically controlled as before. In order to reach the necessary marching speed, the rockets were launched at a slight upward angle. Then, after about 100 meters, the rockets were automatically steered to the line of the targeting optics. According to the guidance of the targeting optics, the antitank rockets were now aimed at the armored vehicle to be struck.

At the beginning of the seventies a start was made in using the BRDM-2 reconnaissance vehicle as a carrier for antitank rockets. In another new move, the monocular targeting optics of the guided-missile gunner was built into the vehicle. This allowed fire control directly from the vehicle.

Finally, the 9P148 launching vehicle was displayed in the November 1977 parade. With this tank destroyer, five 9M111 "Fagot" (Bassoon) or 9M113 "Konkurs" (Competition) missiles could be fired from a folding and lowerable launching rack. It was also possible to place both types of rockets on the launching rack simultaneously. Both rockets have a range of 2,500 meters and are guided semi-automatically.

At the end of 1990, the newest tank destroyer was displayed by the Russian government at a weapons fair in Manila, Philippines. The 9P149 launching vehicle, based on the MT-LB multipurpose towing and transport vehicle, is capable of firing not only the 9M111 and 9M113 antitank missiles, but also the 9M114 "Shturm" (Storm) guided missile. The range of the 9M114 antitank rocket, originally intended to be carried by helicopters, is some 5,000 meters. A warhead with a splinter explosive charge can be used, as well as the hollow-charge warhead.

The 2P26 launcher for four 3M6 "Bumble-bee" (Russian "Shmel") antitank rockets.

---

26. Erhart, K., Panzerabwehrlenkraketen, Berlin 1983, p. 7.

Chapter 9: Tank Destroyers

Firing a wire-guided 3M6 antitank rocket (NATO code: AT-1/SNAPPER).

Launching Vehicle 9P110
Raketnaya samochodno-puskovaya ustanovka 9P110

Made in: USSR
Used in: USSR, CSSR, East Germany, Ethiopia, Hungary, Poland
Developed: 1957-58
Manufactured: 1958 to ca. 1967-68
Crew: 2 men
Fighting weight: 6.0 tons
Length: 5700 mm
Width: 2250 mm
Height: 2800 mm (firing position), 2000 mm (marching)
Armament: 6 launching rails for 9M14 "Malyutka" antitank rockets
Powerplant: 6-cylinder GAZ-40P gasoline engine (90 HP/66 kW)
Top speed: 80 kph (road), 9 kph (water)
Range: 500 km

Launching Vehicle 9P133
Raketnaya samochodno-puskovaya ustanovka 9P113

Made in: USSR
Used in: USSR, Algeria, East Germany, Egypt, Hungary, Iraq, Libya, Morocco, Nicaragua, Poland, Romania, Syria, Yugoslavia
Developed: 1964
Manufactured: 1965 to ca. 1974-75
Crew: 2 men
Fighting weight: 7.20 tons
Length: 5750 mm
Width: 2350 mm
Height: 2800 mm (firing position), 2044 mm (marching)
Armament: 6 launching rails for 9M14 "Malyutka", 9M14M "Malyutka-M" or 9M14P "Malyutka-P" antitank rockets (complete load: 14 rockets)
Powerplant: 8-cylinder GAZ-41 gasoline engine (140 HP/103 kW)
Top speed: 100 kph (road), 10 kph (water)
Range: 750 km

In the first armored troop carriers, called Launching Vehicle 2P27, the launcher had only three rails and the cover folded to the side. It fired wire-guided 3M6 "Bumblebee" rockets.

The 2P32 launcher could launch four 3M11 wire-guided "Closed Front" (Russian "Falanga") antitank rockets (NATO code: AT-2/SWATTER). The launcher cover also folded to the side.

In the 9P110 launcher for six 9M14 "Tiny One" (Russian "Malyutka") antitank rockets, the launcher and cover were raised up together. The picture shows a Soviet naval infantry vehicle.

On the BRDM-2 armored reconnaissance and scouting vehicle there was built the 9P122 launcher for six 9M14 or 9M14M "Little One" antitank rockets.

A scrap yard after the disbanding of the East German People's Army. In the first row are two 9P122 launchers and one 9P133 type (recognizable by the different targeting optics). These vehicles could launch six 9M14, 9M14M or 9M14P (NATO code: AT-3/SAGGER) antitank rockets.

Launch of a 9M14 "Tiny One" antitank rocket (NATO code: AT-3/SAGGER) from the 9P110 launching vehicle.

After the 9P110 launching vehicle was introduced, a BTR-40 armored transporter was also fitted with the launcher for six wire-guided 9M14 antitank rockets.

Chapter 9: Tank Destroyers

Tank destroyers of the Czech Army, seen during training on their 9P122 launchers. Each crew consisted of two men.

The 9P149 launching vehicle can launch the laser-guided 9M114 "Storm-S" (Russian "Schturm-S") antitank rockets (NATO code: AT-6/SPIRAL).

The 9P148 launching vehicle could launch five 9M111 "Bassoon" (Russian, "Fagot" or 9M113 "Competition" (Russian "Konkurs") antitank rockets. In the picture, the outer launcher holds two 9M111 "Bassoon" antitank rockets (NATO code: AT-4/SPIGOT) and the middle one three 9M113 "Competition" rockets (NATO code: AT-5/SPANDREL).

The 9P148 launching vehicle had the launching apparatus on an arm that swung into the interior of the vehicle.

The targeting optics of the 9P148 launcher directed the laser-guided antitank rockets.

The folded and lowered starting apparatus.

10

The Armored Reconnaissance Vehicles

Among the armored reconnaissance vehicles is the PT-76 light armored amphibian. Although it has been used by the fighting forces of the USSR and its allies for many years, it has been replaced more and more since the seventies by special reconnaissance vehicles.

In principle, all units that come into direct contact with the enemy at the front should be involved in reconnaissance. Along with the independent reconnaissance troops, the armored units in particular with tanks and armored personnel carriers needed to form "combat reconnaissance troops." Consideration was also given to forming strengthened and regular units as "reconnaissance units."

The reconnaissance units generally consisted of cooperating tank, engineer and chemical reconnaissance units. Battlefield observers were often included, as well.

Every armored or motorized rifle division had a reconnaissance battalion, with one company for each regiment. They were equipped with the available tanks, rifle and other personnel carriers, and BRM and BRDM armored reconnaissance vehicles. Among others, the PRP-3 and SNAR-10 wheeled armored vehicles were used for battlefield observation.

In every division there was a so-called "long-distance reconnaissance company," which was supposed to operate in the rear of the enemy. It could consist of up to six men, who were landed either by parachute or by helicopter. The reconnaissance vehicle used by this unit was the BRDM armored reconnaissance vehicle, which was adapted from the BTR-40 armored transporter.

The reconnaissance of the air space was up to the troop and regional anti-air defenses. The launching vehicles used for defense against low-flying targets were equipped for the most part with early-warning and altitude-finding radar devices. Reconnaissance and guiding stations were brought into action for mobile launching vehicles in the fighting of aircraft flying at medium and high altitudes.

The artillery divisions, brigades and regiments were supplied not only with armored observation cars, such as the SNAR-10 and 1W17, but with numerous battlefield surveillance, range-finding, night-vision, sound- and light-measuring and radar devices for artillery reconnaissance.

Engineer reconnaissance was carried out by special engineer reconnaissance platoons. They were supplied with armored infan-

The BRDM armored reconnaissance and scouting vehicle. To cross ditches up to 1220 mm wide, it could extend four support wheels between the two axles. The two rear support wheels are lowered here.

In the water, the vehicle was driven by water jets, and could attain a speed of 10 kph.

Chapter 10: The Armored Reconnaissance Vehicles

The bracket for the 7.62 mm SGMT machine gun.

A look at the driver's and commander's area of the BRDM.

try carriers and armored rifle wagons (armored transports). In the mid-eighties, the partial introduction of the IRM engineer reconnaissance vehicle began.

For chemical and nuclear reconnaissance, warning and sensing devices were installed in all command vehicles. There were special vehicles only in the chemical units. As a rule, these were supplied with BRDM-2Ch vehicles. But there were also the BTR-60PB-Ch and BTR-70Ch armored reconnaissance vehicles and the RChM reconnaissance tank.

Armored Reconnaissance and Surveillance Vehicle BRDM (later BRDM-1)
Bronirovanaya rasvedivatelno-dosornaya maschina BRDM)

Made in: USSR
Used in: USSR, Albania, Algeria, Angola, Bulgaria, Congo, Cuba, CSSR, East Germany, Guinea, Mozambique, Romania, Sudan, Zambia, Zimbabwe
Developed: 1956
Manufactured: 1957-1966
Crew: 2 or 3 men
Fighting weight: 5.60 tons
Length: 5600 mm
Width: 2250 mm
Height: 1910 mm
Armament: 1 7.62 mm SGMT machine gun (1250 rounds)
Powerplant: 6-cylinder GAZ-40P gasoline engine (90 HP/66kW)
Top speed: 80 kph (road), 9 kph (water)
Range: 500 km

In the rear of the vehicle there was enough space to carry up to six observers. This was why the vehicle was falsely called an armored troop carrier in some countries, though not used as such.

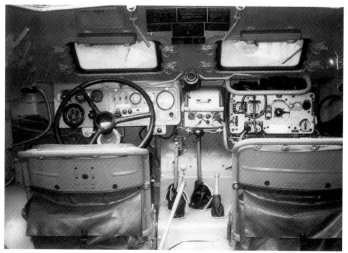

The driver's and commander's seats and controls, plus the R-123 tank radio set.

99

The BRDM armored reconnaissance and scouting vehicle as SPW-40P armored troop carrier crossing an obstacle in East German Army service.

The BRDM-2 armored reconnaissance and scouting vehicle of an East German Army unit, as SPW-40P2, during the October 1980 parade in Berlin.

Armored Reconnaissance and Surveillance Vehicle BRDM-2
Bronivanaya rasvedivatelno-dozornaya maschina BRDM-2

Made in: USSR/CIS
Used in: USSR/CIS/Russia, Afghanistan, Algeria, Angola, Benin, Botswana, Bulgaria, Cape Verde Islands, Central African Republic, Chad, Congo, Cuba, Djibouti, East Germany, Egypt, Equatorial Guinea, Ethiopia, Guinea, Guinea-Bissau, Hungary, India, Iraq, Israel, Libya, Mauretania, Mongolia, Morocco, Mozambique, Nicaragua, Peru, Poland, Romania, Sao Tome, Seychelles, Somalia, South Yemen, Sudan, Syria, Tanzania, Vietnam, Yugoslavia, Zambia, Zimbabwe
Developed: 1962
Manufactured: 1963-1989
Crew: 3 men
Fighting weight: 7.0 tons
Length: 5750 mm
Width: 2350 mm
Height: 2310 mm
Armament: 1 14.5 mm KPWT heavy machine gun (500 rounds)
          1 7.62 mm SGMT or PKT machine gun (2000 rounds)
Powerplant: 8-cylinder GAZ-41 gasoline engine (140 HP/103 kW)
Range: 750 km

Engineer Reconnaissance Vehicle IRM
Inshenernaya rasvedivatelnaya maschina IMR

Made in: USSR
Used in: USSR/CIS/Russia
Developed: ca. 1980
Manufactured: since 1982-83
Crew: 2 men
Fighting weight: 17.0 tons
Length: 8220 mm
Width: 3150 mm
Height: 2400 mm
Armament: 1 7.62 mm PKT-M machine gun
Powerplant: 6-cylinder Diesel engine (300 HP/220 kW)
Top speed: 100 kph (road), 10 kph (water)
Range: 400 km

A BRDM-1Ch chemical reconnaissance vehicle.

A collection of special vehicles based on the BRDM-2 armored reconnaissance and scouting vehicle. In front is a BRDM-2U command vehicle, behind it 9P122 and 9P133 launchers, and at right a BRDM-2 and BRDM-2Ch chemical reconnaissance vehicles.

Chapter 10: The Armored Reconnaissance Vehicles

A BRDM-2Ch chemical reconnaissance vehicle.

Ground Reconnaissance Station of the Artillery SNAR-10
Stanziya nasemnoy artilleriyskoy rasvedki SNAR-10

Made in: USSR
Used in: USSR/CIS/Russia, East Germany
Developed: ca. 1970-71
Manufactured: since 1971-72
Crew: 4 men
Gighting weight: 12.2 tons
Length: 6454 mm
Width: 2850 mm
Height: 2390 mm (marching)
Armament: 1 7.62 mm PKT machine gun
Powerplant: 8-cylinder JaMZ-238W Diesel engine (240 HP/177 kW)
Top speed: 61.5 kph
Range: 500 km

The BRDM-2Ch (here as SPW-40P2Ch in the East German Army) was armed with a 14.5 mm KPWT heavy machine gun and a 7.62 mm PKT machine gun.

Some BRDM-2Ch were armed with just one 7.62 mm machine gun.

The BTR-70Ch chemical reconnaissance vehicle.

The RChM chemical reconnaissance vehicle.

The IMR engineer reconnaissance vehicle.

The BRM-1K armored reconnaissance vehicle with raised antenna for the 1RL133-1 reconnaissance station (NATO code: TALL MIKE).

The turret rear of a BRM-1K, with the 1RL133-1 reconnaissance radio antenna folded down.

The BRM-1K (Russian Boyevaya rasvedivatelnaya maschina) armored reconnaissance car.

Chapter 10: The Armored Reconnaissance Vehicles

Over the two rear entries, which were also used to hold fuel containers, the semi-telescopic antenna mast was attached, seen here in marching position.

The BRM-1K with its doors open, revealing the drivers' seats.

The BRM-1K with raised semi-telescopic antenna.

The BRM-3K armored reconnaissance vehicle with 30 mm 2A72 machine cannon, laser range finder, laser target light, remaining-light observation device and battlefield surveillance radar.

The PRP-3 mobile reconnaissance post (Russian Podvishniy rasvedivatelniy post) with folding antenna for reconnaissance station (NATO code: SMALL FRED). The vehicle was armed with a 7.62 mm PKT-M tank machine gun.

The SNAR-2 artillery ground reconnaissance station (NATO code: PORK TROUGH) on the AT-L artillery tractor.

The SNAR-10 artillery ground reconnaissance station (Russian Stantziya nasemnoi artilleriyskoi rasvedki) on the MT-LB multipurpose towing and transport vehicle, with the 1RL-232 artillery radio reconnaissance station (NATO code: BIG FRED).

The PPRU-1 mobile reconnaissance and command post (NATO code: DOG EAR) of the troop air defense, on the MT-LBu multipurpose towing and transport vehicle.

The P-40 surveillance station on basic vehicle 426U (NATO code: LONG TRACK) of the air defense units. Six-man crew, weight 36.5 tons, length 7300 mm, width 3100 mm, height 3900 mm (antenna in marching position), powered by a 12-cylinder W-401G Diesel engine (465 HP/342 kW), top speed 55 kph.

Chapter 10: The Armored Reconnaissance Vehicles

The 1S91M reconnaissance and guiding station (NATO code: STRAIGHT FLUSH) of the 2K12 "Cube" (Russian "Kub") AA rocket complex.

The 9S15 and (right) 9S32 rocket guiding stations of the S-300W AA rocket complex.

The 1S32 rocket guiding station (NATO code: PAT HAND) of the 2K11 "Circle" (Russian "Krug") AA rocket complex.

11

Armored Transporters and Combat Vehicles of the Infantry

During World War II, the degree of mechanism in the Red Army increased extremely. The more the possible attacking speed of the armored divisions increased, the more necessary did the equipping of the rifle divisions with motorized vehicles become in order to keep up. In the process of the steadily growing numbers of motor vehicles, it became clearer that for "penetration into the depths of the enemy's defenses" the riflemen had to be equipped with armored vehicles with off-road capability. Thus, in the first postwar years an extensive developmental program began, to create the armored troop carrier (below: BTR = "Bronetransporter, armored transporter).

Production of the BTR-40 and BTR-152 began as early as 1950. For the BTR-40, the chassis of the GAZ-63 light off-road (4x4) truck, fitted with a lightly armored body, was used. It was intended to transport up to eight fully equipped riflemen onto the battlefield. Its armament consisted of a 7.62 mm SGMT tank machine gun on an open mount. Along with its actual designation as a troop carrier, the BTR-40 was used particularly for reconnaissance, commando, intelligence and other special tasks. In terms of design, the vehicle did not impose any particular demands in driving and maintenance, since it was almost completely identical to the GAZ-63 truck. On the other hand, its off-road capability was seen as not sufficient. In the vehicles used for reconnaissance in particular, it was seen that every water obstacle that had no bridge could be crossed up to a depth of 0.9 meters. In order to protect the riflemen from aircraft weapons, the use of a closed body was seen as necessary around 1957. The vehicles of the last series of BTR-40B built in 1957 finally had armor plate on top.

The development of a transport vehicle for use in the rifle divisions and capable of carrying up to 17 men began as early as 1948. It too was built on a truck chassis. The BTR-152 consisted basically of the ZIL-151 (6x6) medium truck with an armored body. During its production it was tried, by introducing a tire-pressure regulator, to improve its off-road capability. In the BTR-152W model introduced in 1955, the tubing was still attached outisde at first. This was particularly sensitive during off-road driving and crossing barriers. Two years later, in the BTR-152W1, it was possible to set the air-pressure tubes in the axle ends, so that the danger of external damage could be strongly reduced. As in the BTR-40, the BTR-152K built from 1958-59 on was also given an enclosed, armored body at last.

Although it was actually developed as a troop transporter, the BTR-40, as already noted, was used as a reconnaissance vehicle. Its successor was finally built only as an armored reconnaissance and scouting vehicle.

While tanks with appropriate wading equipment could cross water barriers with depths up to four meters, the BTR-152 was directed to artificial crossings when the water depth was more than 0.8 meters. In order to make up for this weakness, experiments on a new troop carrier concentrated on the amphibious character of these vehicles. In 1959 a six-wheel ZIL-153 transporter was proposed, but it was not introduced. One year later the BTR-60P was finally ready for introduction. The letter "P" in its designation (for "plava-ushiy," amphibian) indicated its ability to move in water.

In 1950-60 the military theoreticians who regarded warplanes as too fast for action against armored vehicles were finally able to prevail. Just as the anti-aircraft machine gun was not mounted on tanks, the troop transporters were produced again with open tops. Here, too, in conjunction with the introduction of antitank helicopters, top armor for the vehicles of the motorized rifle divisions was regarded as a necessity. The version with enclosing armor, built from 1963 to 1966, was used by the Soviet Army as BTR-60PA. To protect the operators of the top-mounted 12.7 mm heavy machine gun from the effects of weapons and splinters, tests were carried out on the wheeled test vehicle "Object 1200" for a turning turret armoed with machine guns. After examining the test results, though, changing the chassis was ruled out. Instead, the BTR-60PA was now built in large numbers with a turret for a 14.5 mm KPWT heavy machine gun and a 7.62 mm SGMT machine gun; this was built from 1965 to 1976 as the BTR-60PB.

In the following years, numerous experiments were conducted to improve the eight-wheel transporter's fighting and driving qualities. Among others, it was suggested in 1965 to use tracks in place of the two middle axles. In 1970-71 the possibility of using the BMP-1 turret on the BTR-60 was tested. Although the armor protection was improved by setting more armor plate at an angle, the only real improvement to the BTR-70 (1972 model) consisted of the two side exit doors for the motorized riflemen. The Russian

Chapter 11: Armored Transporters and Combat Vehicles of the Infantry

armament industry stuck with the use of two gasoline engines, the GAZ-49 and GAZ-49B, much too long. To be sure, the GAZ engines were extremely robust, but in tandem they quickly caused overheating, an experience that the Red Army had already had with the SU-76 self-propelled gun in World War II. Only in 1982 did the Soviet designers decide to turn away from this principle. Now the use of gasoline engines was given up, and only a 260 HP Diesel engine was used. In order to be able to use the machine guns in the turning turret against low-flying air targets, their upward elevation was expanded considerably. Some BTR-70s that were still being built when production of the BTR-80 began were already equipped with the BTR-80's turret.

After the development of an amphibian reconnaissance tank was concluded, there appeared on the drawing boards at the Volgograd Tank Works the first conceptions using the chassis of the PT-76 amphibian tank for a tracked amphibian armored transporter. After three years' work, the first rifle regiments of the armored divisions were finally supplied with the BTR-50P armored transporter.

This troop transport was also built at first without armor on top of the transport area. As of 1958, this vehicle too was built with a closed body. This version was used by the forces of the USSR and its allies as BTR-50PK (1958 model).

On November 7, 1967, a completely new type of vehicle for motorized riflemen was first shown at the Moscow parade. Along with its relatively large-caliber weapon and its high speed, its very low silhouette aroused astonishment. The introduction of this vehicle, with the designation "Infantry Combat Vehicle BMP" had already begun a year before. The inclusion of a launching apparatus for antitank missiles was also new.

Although it became known internationally only after the publishing of a photo taken in Afghanistan, the designers had long been working on the successor model to the BMP. With the use of multi-layered armor plate, the use of the 73 mm 2A28 gun was no longer regarded as sufficient. Since the recoilless gun was only capable of firing post-accelerated fin-stabilized missiles of relatively low speed, only hollow-charge and splinter-explosive shells could be used. In 1972, various machine cannons were tested for installation in the BMP turret. The decision was finally made in favor of the 30 mm 2A42 machine cannon, with both a high rate of fire and a penetrating power sufficient to deal with the armor of all combat vehicles except tanks. Through its high arc of elevation and second targeting optics, this weapon could also be used against low-flying aircraft. For action against tanks, the possibility of launching antitank guided missiles was retained. The infantry took over the new combat vehicle as BMP-2 (1974 model). After 1981 the BMP-2 began to be fitted with additional armor. This reached the troops as BMP-2D (1981 model).

Under the pressure of a 1.5-ton increase in weight, the BMP-2's track aprons had to be made into floats. Since a further increase in weight was to be avoided, tests were made since the late seventies to develop a new tracked vehicle for the motorized rifle units. One of the suggestions was developed into a prototype as "Object 688" (1981 model).

In the mid-eighties, the first BMP-3 infantry combat vehicles were finally received by the Russian armed forces. The troop carrier is built in the Kurgan machine shop. It was first shown publicly at the 1990 Moscow parade. The need to retain its amphibian capability led to a compact design with high sidewalls. The number of fixed weapons was increased. In the turning turret is a 100 mm 2A70 gun, which fires 9M117 guided missiles of the 9K116 "Bastion" guided-missile complex. The improved 30 mm 2A72 machine cannon was installed beside it. In addition, two 7.62 mm machine guns in ball mantlets are mounted on the left and right sides of the upper front of the hull, each one operated by a gunner behind it. Since there are also two gunners in the front of the hull along with the driver, the motor was located in the right rear part of the vehicle. For water travel, a water-jet drive with two folding openings was installed in the rear wall of the vehicle.

Armored Transporter BTR-40, BTR-40W, BTR-40K
Bronetransporter BTR-40 (obr. 1950 g.), BTR-40W (obr. 1956 g.), BTR-40K (obr. 1957 g.)

Made in: USSR (auto factory in Gorki)
Used in: USSR, Afghanistan, Albania, Bulgaria, Burundi, Cambodia, China, Cuba, East Germany, Egypt, Equatorial Guinea, Ethiopia, Guinea, Guinea-Bissau, Indonesia, Iran, Laos, Mozambique, Nicaragua, North Korea, North Yemen, Solamia, South Yemen, Syria, Tanzania, Uganda, Vietnam, Yugoslavia, Zaire
Developed: 1949
Manufactured: 1950-1958
Crew: 2 + 8 men
Fighting weight: 5.30 tons
Length: 5000 mm
Width: 1900 mm
Height: 1930 mm (BTR-40 and 40W), 2060 mm (BTR-40K)
Armament: 1 7.62 mm SGMT machine gun (1250 rounds)
Powerplant: 6-cylinder GAZ-40 gasoline engine (78 HP/57 kW)
Top speed: 78 kph
Range: 285 km

For lack of suitable armored vehicles for the infantry, many BA-64 armored cars were used without turrets as armored transports.

The BTR-40 armored transporter (1950 model), armed with a 7.62 mm SGMT tank machine gun, seen with an East German Army unit on March 12, 1961.

The driver's seat in the BTR-40. There were also BTR-40s with external tire-pressure regulators. In these vehicles, designated BTR-40W (1956 model), the release valves were between the two seats.

Armored Transporter BTR-152, BTR-152W, BTR-152W1, BTR-152K

Bronetransporter BTR-152 (obr. 1952 g.), BTR-152W (obr. 1955 g.), BTR-152W1 (obr. 1957 g.), BTR-152K (obr. 1959 g.)

Made in: USSR (auto factory in Miass)
Used in: USSR, Afghanistan, Albania, Algeria, Angola, Bulgaria, Cambodia, Central Africa, China, Congo, Cuba, East Germany, Egypt, Ethiopia, Guinea, Guinea-Bissau, Indonesia, Iran, Iraq, Laos, Mali, Mongolia, Mozambique, Nicaragua, North Korea, North Yemen, Poland, Romania, Somalia, South Yemen, Sri Lanka, Sudan, Syria, Tanzania, Vietnam, Yugoslavia, Zimbabwe
Developed: 1949
Manufactured: 1950-1959
Crew: 2+17 men (BTR-152, 152W and W1), 2+13 men (BTR-152K)
Fighting weight: 8.60 tons (BTR-152), 9.0 tons (BTR-152W), 8.95 tons (BTR-152W1 and K)
Length: 6830 mm (with winch), 6550 mm (without winch)
Width: 2320 mm
Height: 2360 mm (BTR-152), 2410 mm (BTR-152W and W1), 2710 mm (BTR-152K)
Armament: 1 7.62 mm SGMT machine gun (1250 rounds)
Powerplant: 6-cylinder ZIL-123 gasoline engine (110 HP/81 kW)
Top speed: 65 kph
Range: 550 km

The BTR-40B (1957 model) armored transporter with closed rear body.

The BTR-152 (1950 model) armored transporter.

Chapter 11: Armored Transporters and Combat Vehicles of the Infantry

Side view of the BTR-152 (1950 model) armored transporter.

The BTR-152W (1955 model) armored transporter was equipped with an external tire-pressure regulator. This one carries a ZPTU-2 self-propelled AA gun mount armed with 14.5 mm heavy twin machine guns.

Armored Transporter BTR-60P, BTR-60PA, BTR-60PB
Bronetransporter BTR-60P (obr. 1960 g.), BTR-60PA (obr. 1963 g.), BTR-60PB (obr. 1965 g.)

Made in: USSR (auto factories in Gorki and Ulyanovsk)
Used in: USSR/CIS/Russia, Afghanistan, Algeria, Angola, Botswana, Bulgaria, Cambodia, China, Djibouti, East Germany, Ethiopia, Guinea, Guinea-Bissau, India, Iran, Iraq, Libya, Mali, Mongolia, Mozambique, Nicaragua, North Korea, North Yemen, Romania, Somalia, South Yemen, Syria, Vietnam, Yugoslavia, Zambia
Developed: 1959-60
Manufactured: 1960-1976
Crew: 2+14 men (BTR-60P), 2+12 men (BTR-60PA), 2+8 men (BTR-60PB)
Fighting weight: 9.80 tons (BTR-60P and PA), 10.3 tons (BTR-60PB)
Length: 7220 mm
Width: 2906 mm
Height: 2055 mm (BTR-60P and PA), 2375 mm (BTR-60PB)
Armament: 1 7.62 mm SGMT machine gun (1250 rounds) (BTR-60P and PA)
and 1 14.5 mm KPWT heavy machine gun (500 rounds)
 1 7.62 mm SGMT or PKT machine gun (2000 rounds) (BTR-60PB)
Powerplant: 2 6-cylinder GAZ-49B gasoline engines (2x90 HP/ 2x66 kW)
Top speed: 80 kph (road), 10 kph (water)
Range: 500 km

In the BTR-152W1 armored transporter (1957 model), tire-pressure regulation was done by the air lines in the axle ends.

The BTR-152W1 (1957 model) armored transporter.

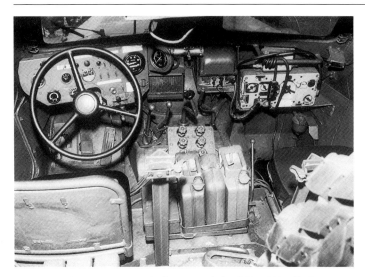

The driver's and commander's seats of the BTR-152 (1957 model). Between the seats in front of the detoxification set are the release valves of the tire-pressure regulator.

The BTR-152K (1959 model) armored transporter had a closed top. In the same year, the ZIL-153 armored transport (1959 model) was created—a three-axle prototype that much resembled the successor BTR-60 externally.

Armored Transporter BTR-70
Bronetransporter BTR-70 (obr. 1972 g.)

Made in: USSR (auto factories in Gorki and Ulyanovsk)
Used in: USSR/CIS/Russia, East Germany
Developed: 1971
Manufactured: 1972-1985
Crew: 3+7 men
Fighting weight: 12.0 tons
Length: 7510 mm
Width: 2790 mm
Height: 2235 mm
Armament: 1 14,5 mm KPWT heavy machine gun (500 rounds)
            1 7.62 mm PKT machine gun (2000 rounds)
Powerplant: 2 8-cylinder ZMZ-49-05 gasoline engines (2x115 HP/ 2x85 kW)
Top speed: 80 kph (road), 10 kph (water)
Range: 400 km

Armored Transporter BTR-80
Bronetransporter BTR-80 (obr. 1982 g.)

Made in: USSR/CIS/Russia (auto factories in Gorki and Ulyanovsk)
Used in: USSR/CIS/Russia
Developed: 1982-83
Manufactured: since 1984
Crew: 3+7 men
Fighting weight: 13.6 tons
Length: 7650 mm
Width: 2900 mm
Height: 2350 mm
Armament: 1 x 14.5 mm KPWT heavy machine gun (500 rounds)
            1 7.62 mm PKT-M machine gun (2000 rounds)
Powerplant: 8-cylinder (multifuel) Diesel engine (260 HP/191 kW)
Top speed: 80 kph (road), 10 kph (water)
Range: 600 km

The BTR-60P armored transporter (1960 model).

Although the troop carrier had already been built with a closed top in 1959, the first eight-wheel-drive armored troop carriers were built with an open fighting compartment.

Chapter 11: Armored Transporters and Combat Vehicles of the Infantry

The BTR-60PA armored troop carrier (1963 model) now had a closed hull roof again.

A BTR-60PA armored transporter (1963 model) of a Soviet naval infantry unit.

A motorized rifle platoon of the Soviet Army on the attack.

Armored Transporter BTR-50P, BTR-50PA, BTR-50PK
Bronetransporter BTR-50P (obr. 1954 g.), BTR-50PA (obr. 1954 g.), BTR-50PK (obr. 1958 g.)

Made in: USSR (tank factory in Volgograd)
Used in: USSR, Afghanistan, Albania, Algeria, Angola, Bulgaria, China, Congo, CSSR, Cyprus, East Germany, Egypt, Finland,. Guinea, India, Iran, Iraq, Israel, Libya, Morocco, Nicaragua, North Korea, Poland, Romania, Somalia, Syria, Vietnam, Yugoslavia
Developed: 1953-54 (BTR-50P), 1958 (BTR-50PK)
Manufactured: 1954 to 1969-70
Crew: 3+18 men
Fighting weight: 13.0 tons (BTR-50P), 14.2 tons (BTR-50PK)
Length: 7000 mm (BTR-50P), 7070 mm (BTR-50PK)
Width: 3140 mm
Height: 2030 mm
Armament: 1 7.62 mm SGMT machine gun (1250 rounds)
Powerplant: 6-cylinder W-6 Diesel engine (240 HP/177 kW)
Top speed: 44.6 kph (road), 10.2 kph (water)
Range: 250 km

The BTR-60PA (1963 model) was regarded as an armored troop carrier in the East German Army.

The exhaust pipe of the left-mounted six-cylinder GAZ-49B gasoline engine, which gave 90 HP (66 kW).

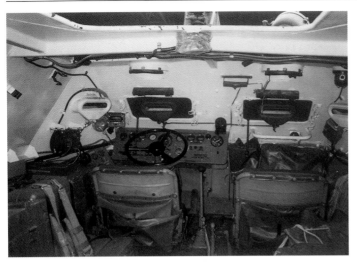
The driver's and commander's seats of the BTR-60PA (1963 model).

BTR-60PB armored transporters (1966 model) on the move.

The BTR-60PB armored transporter (1966 model).

The BTR-80 armored transporter (1982 model).

The BTR-60PB armored transporter (1966 model) of a Soviet naval infantry unit landing on the Baltic coast.

The BTR-70 armored transporter (1972 model), with the right side entry hatch opened.

Chapter 11: Armored Transporters and Combat Vehicles of the Infantry

The BTR-60PB armored transporter (1966 model) of the East German secret service chief.

Infantry Combat Vehicle BMP-1, BMP-1P
Boewaya maschina pechoti BMP-1 (obr. 1966 g.)

Made in: USSR (tank factory in Ishewsk), CSSR
Used in: USSR/CIS/Russia, Afghanistan, Algeria, Bulgaria, CSSR, Cuba, East Germany, Egypt, Ethiopia, Finland, Hungary, India, Iran, Iraq, Kuwait, Libya, Mongolia, North Korea, North Yemen, Poland, Syria, Yugoslavia
Developed: 1966
Manufactured: 1966-1982
Crew: 3+8 men
Fighting weight: 13.5 tons
Length: 6735 mm
Width: 2940 mm
Height: 2068 mm
Armament: 1 73 mm 2A28 cannon (40 rounds)
 1 7.62 mm PKT machine gun (2000 rounds)
 1 9S428 launching apparatus for antitank guided missiles (BMP-1 = 4 PALR 9M14M "Malyutka", BMP-1P = 4 PALR 9M111 "Fagot" or 9M113 "Konkurs")
Powerplant: 6-cylinder UTD-20 Diesel engine (295 HP/217 kW)
Top speed: 65 kph (road), 7 kph (water)
Range: 600 km

Infantry Combat vehicle BMP-2, BMP-2D
(Boewaya maschina pechoti BMP-2 (obr. 1974 g.), BMP-2D (obr. 1981 g.)

Made in: USSR, CSSR (tank factory in Ishewsk)
Used in: USSR/CIS/Russia, Algeria, CSSR, East Germany, Finland, India, Iraq, Kuwait
Developed: 1974
Manufactured: 1974-1985
Crew: 3+7 men
Fighting weight: 13.8 tons (BMP-2), 14.5 tons (BMP-2D)
Length: 7295 mm (gun rear mount)/6735 (gun forward mount)
Width: 3140 mm
Height: 2365 mm
Armament: 1 30 mm 2A42 cannon (500 rounds)
 1 7.62 mm PKT machine gun (2000 rounds)
 1 launching apparatus for antitank missiles (4 PALR 9M111 "Fagot" or 9M113 "Konkurs")
Powerplant: 6-cylinder UTD-20 Diesel engine (295 HP/217 kW)
Top speed: 65 kph (road), 7 kph (water)
Range: 550 km

The BTR-70 armored transporter (1972 model)

The BTR-80 armored transporter (1982 model).

Although problems with synchronizing the two gasoline engines had existed constantly for decades, only in the BTR-80 (1982 model) was the tandem principle given up and a 260 HP Diesel engine used.

Infantry Combat Vehicle BMP-3
Boewaya maschina pechoti BMP-3

Made in: USSR/CIS/Russia (Kurgan machine factory)
Used in: USSR/CIS/Russia
Developed: 1982-83
Manufactured: 1984-85 to date
Crew: 3+7 men
Fighting weight: 18.7 tons
Length: 6725 mm
Width: 3300 mm
Height: 2450 mm
Armament: 1 100 mm 2A70 cannon (22 rounds)
　　　　　　1 30 mm 2A72 cannon (500 rounds)
　　　　　　2 7.62 mm PKT-M machine guns (6000 rounds)
Powerplant: 10-cylinder (multifuel) Diesel engine UTD-29M (500 HP/368 kW)
Top speed: 70 kph (road), 20 kph (water)
Range: 600 km

The first fully tracked vehicle of the motorized rifle regiments was the BTR-50P armored transport (1954 model). This vehicle also had an open top at first.

Armed with the 14.5 mm KPWT heavy machine gun was the BTR-50PA tracked amphibious armored transport (1954 model).

As of 1958 the motorized rifle units received the closed-top BTR-50PK armored transport (1958 model). This one was photographed in April 1960.

Chapter 11: Armored Transporters and Combat Vehicles of the Infantry

The BTR-50PK (1958 model) crossing trenches in August 1969.

The BMP-1 infantry combat vehicle (1966 model).

Over the 73 mm 2A28 gun was the firing apparatus for the wire-guided 9M14 or 9M14M "Tiny One" (Russian "Malyutka") antitank rocket (NATO code: AT-3/SAGGER)

The BMP-1 infantry combat vehicle (1966 model) in winter maneuvers.

The BMP-1 infantry combat vehicle could reach a very high speed, even in rough terrain.

The BMP-1 infantry combat vehicle running through various terrains and obstacles.

Setting the antitank gun on the launching rail was done through an opening hatch above the 73 mm gun, but the gun had to be elevated as high as possible.

The BMP-1 was driven by its tracks in water.

The author driving the BMP-1 infantry combat vehicle (1966 model).

For the BMP-1 to be able to launch the 9M111 "Bassoon" (Russian, "Fagot") (NATO code: AT-4/SPIGOT) and 9M113 "Competition" (Russian "Konkurs") (NATO code: AT-5/SPANDREL), a brace was added on the right side of the turret roof to attach the 9P135 or 9P135M launcher.

Chapter 11: Armored Transporters and Combat Vehicles of the Infantry

The BMP-1P infantry combat vehicle with the 9P135 launcher and 9M111 "Bassoon" antitank rocket.

The BMP-1 infantry combat vehicle in a Mongolian army unit, re-equipped with a bracket to hold the 902 foglaying apparatus.

The BMP-2 infantry combat vehicle (1974 model).

On the rear part of the turret roof, the 9P135M launcher for the 9M111 or 9M113 antitank rockets could be attached.

117

The BMP-2 infantry combat vehicle, known as the BMP-2 armored troop carrier in the East German Army, loaded on a Czech P50/80 low loader trailer.

The BMP-2D infantry combat vehicle (1981 model) first appeared in Afghanistan.

To give the BMP-2 infantry combat vehicles better protection from hollow-charge ammunition, at least on the sides, they were fitted with additional armor.

An attack by a motorized rifle company, equipped with BMP-2 vehicles and supported from the air by Mi-24 helicopters (NATO code: HIND E/F).

Chapter 11: Armored Transporters and Combat Vehicles of the Infantry

The BMP-3 infantry combat vehicle.

The BMP-3 is armed with a 100 mm 2A70 gun and a parallel-axis 2A72 30 mm gun in its turret, plus two 7.62 mm PKT-M tank machine guns in the bow.

12

Special Armored Vehicles of the Soviet Army, its Allies and the Armies that Evolved after the Dissolving of the USSR

For all special tasks in the realms of service, repair, rescue, overcoming obstacles, removing barrages and building positions, special vehicles were built on the basis of existing armored vehicles. The armored command vehicles likewise belong among the special types. Until the beginning of the fifties, the Soviet leadership scarcely took the trouble to equip their combat units with command tanks. In 1953 it had to be estimated that the means of intelligence and forms of organization were not equal to the tasks. With the equipment at hand, commanders at all levels found themselves unable to react quickly to unforeseeable and troop-threatening situation changes. It was requested that the latest scientific knowledge be applied to the realms of radio electronics, automation and calculation to create the appropriate technical basis for troop command. Only in 1958, though, could means and systems of communication, that would allow fast transmission of large numbers of telegraphic and telephonic information, begin to be introduced. The new models had several intelligence channels, higher performance, better protection from disturbance, less weight and smaller dimensions. In the course of reorganizing intelligence methods, the leaders of smaller units were now equipped with mobile radio stations. Commanders of the armored and motorized rifle divisions in particular, from company chiefs up, were supplied with armored radio vehicles. The tanks of company chiefs were equipped with a second radio set. The tanks of the battalion commanders received not only a second radio, but also a semi-telescopic mast ten meters long. The range could thus be increased to 70 kW. In order to be able to install the second radio in the tanks on hand, the ammunition racks in the back of the turret had to be removed. While the principle of command from ahead had already become customary in nearly all of the world's armies in World War II, the soviet Army began to equip regimental commanders with command tanks only in the sixties. Along with the second radio and semi-telescopic mast, a generator was installed on the bulkhead of the engine room. In the mid-sxities a nuclear radiation warning and spotting device was added. All command tanks received the suffix "K" ("Kommanda," command) or "U" ("Upravlenie," leading) to their designations.

Along with the battle tanks, armored troop carriers were also equipped as command tanks. In the following years, the commanders at all levels of command in the ground combat troops were supplied with armored command vehicles, which included

on tanks: T-34-85K, T-54MK, T-55AK, T-55AMK, T-62K, T-64AK, T-64B1K, T-64B1KW, T-72K, T-72AK, T-72BK, T-80BWK

on reconnaissance tanks: BRM-1K, BRDM-U, BRDM-2U

on troop transports: BTR-50PU, BMP-1K, BMP-1PK, BMP-1KSch, MP-31, BMP-2K

on the BTR-152 armored troop carrier: BTR-152U, R-118, R-118AMZ, R-118BM

on the BTR-60PA/PB armored troop carrier: BTR-60PBK, P-238BT, P-240BT, P-241BT, PU-12, R-137, R-137B, R-145BM, R-150, R-156, R-396, R-409BM, R-419BR, R-440B, R-975M, R-975M1

on the BTR-70 armored troop carrier: KSchM, MBP

on the MT-LB transport tank: R-378, R-421B

on the MT-LBu command vehicle: R-330P, MP-21R, MP-22, MP-22R, MP-23, MP-23R, MP-24R, MP-24R2, MP-24M, MP-25, MP-25R, MP-95, 9S714N-18, KSchM-9S743, 9W514, R-161B, R-409BM, KDChR-1N

on air-landed tanks: BMD-1K, BMD-1PK, BMD-1R, BMD-1KSch, R-440ODB

The T-55AK command tank. A watertight container on the left side of the turret held the 16-meter semi-telescopic mast.

Chapter 11: Special Armored Vehicles of the Soviet Army

The T-55AK command tank.

Command Vehicle BRDM-2U
Bronirovanaya razvedyvatelno-gozornaya maschina BRDM-2Upravlenie

Made in: USSR/CIS
Used in: USSR/CIS/Russia, East Germany
Developed: 1964-65
Manufactured: 1965-66 to 1989
Crew: 4 men
Fighting weight: 6.50 tons
Length: 5750 mm
Width: 2350 mm
Height: 2470 mm
Powerplant: 8-cylinder GAZ-41 gasoline engine (140 HP/103 kW)
Top speed: 100 kph (road), 10 kph (water)
Range: 750 km

Command-Staff Leadership Car BTR-50PU
Komandno-stabnaya maschina upravleniya BTR-50PU

Made in: USSR
Used in: USSR, East Germany
Developed: 1958
Manufactured: 1958-1970
Crew: 3+7 men
Fighting weight: 14.3 tons
Length: 7070 mm
Width: 3140 mm
Height: 2050 mm
Powerplant: 6-cylinder W-6 Diesel engine (240 HP/177 kW)
Top speed: 44.6 kph (road), 10 kph (water)
Range: 250 km

The T-55AK command tank of company chiefs and battalion commanders had a second radio. To use both radios simultaneously with one antenna, an antenna variometer was mounted between them.

The attachment of the 16-meter semi-telescopic mast on a regimental commander's T-55AK.

121

The T-64AK command tank used by company chiefs and battalion commanders.

The T-64AK command tank. These command vehicles based on the T-64 medium tank had no AA machine gun.

Command-Staff Car BMP-1KSch

Komandno-stabnaya maschina BMP-1KSch

Made in: USSR/CIS/Russia
Used in: USSR/CIS/Russia
Developed: 1972
Manufactured: 1976 to date
Crew: 3 men
Fighting weight: 13.0 tons
Length: 6735 mm
Width: 2940 mm
Height: 2068 mm
Armament: 1 7.62 mm PKT tank machine gun (2000 rounds)
Powerplant: 6-cylinder UTD-20 Diesel engine (295 HP/217 kW)
Top speed: 65 kph (road), 7 kph (water)
Range: 550 km

Command-Staff Car R-145BM

Komandno-stabnaya maschina R-145BM

Made in: USSR
Used in: USSR/CIS/Russia, East Germany
Developed: 1966-67
Manufactured: ca. 1968-1976
Crew: 3+3 men
Fighting weight: 10.5 tons
Length: 7220 mm
Width: 2825 mm
Height: 2055 mm (without body), 3360 mm (with body, mast retracted)
Powerplant: 2 6-cylinder GAZ-49B gasoline engines (2x90 HP/2x66 kW)
Top speed: 80 kph (road), 10 kph (water)
Range: 400 km

The BTR-50PU command-staff vehicle (Russian Komandno-stabnaya maschina uprawleniya).

The BTR-50PU command-staff vehicle during Warsaw Pact "Quartette" maneuvers, September 9-14, 1963.

Chapter 11: Special Armored Vehicles of the Soviet Army

The BMP-1K command vehicle with the KMT-8 mine removal device.

The BMP-1KSch command-staff vehicle (1972 model) with its telescopic antenna in marching position.

The KSchM-R-80 or KSchM-R-81 command-staff vehicle was used to command the tank-destroyer units equipped with the MT-LB multipurpose towing and transport vehicle.

The raised telescopic antennas of two BMP-1KSch command-staff vehicles. Although developed in 1972, their production began only in 1976.

The 1W13 command-staff vehicle on the MT-LBu multipurpose towing and transport vehicle, armed with the 12.7 mm DSchK-M AA machine gun (1938-46 model).

The 1W14 command-staff vehicle of a battery chief. Among its armament was a PKMB 7.62 mm machine gun that could be used against aircraft.

The 1W15 command-staff vehicle of a battalion commander in an artillery unit. It was also armed with the 7.62 mm PKMB machine gun.

The chief of staff of an artillery battery used the 1W16 command-staff vehicle. The DSchK-M AA machine gun (1938-46 model) could be mounted on this armored vehicle.

Rear view of the 1W16 command-staff vehicle.

An artillery battery with 1W13, 1W14, 1W15 and 1W16 command-staff vehicles, 2-S1 "Carnation" self-propelled artillery guns, MT-LB multipurpose towing and transport vehicles towing 100 mm T-12 anti-tank guns, 9P31 launchers, ZSU-23-4W1 "Shilka" self-propelled AA guns, PU-12 and BRDM-2U command vehicles.

The KSchM-9S743 command-staff vehicle with R-330P VHF interference set and generator at the rear, seen in Afghanistan.

Chapter 11: Special Armored Vehicles of the Soviet Army

The BRDM-U command vehicle with its telescopic antenna raised.

The BRDM-2U command vehicle, unlike the BRDM armored reconnaissance and scouting vehicle, had no turret. Instead, a box was mounted on the rear to carry a generator during action.

The BRDM-2 could also be used as the BRDM-2M/F courier vehicle (Russian Feldegerskii BRDM-2).

The BRDM-2U command vehicle with turret and frame antenna for the R-113 radio.

BTR-152U of the Russian Military Police.

The KSchM command-staff vehicle was based on the BTR-60PA armored transport.

The BTR-152U was used by the East German Army as the SPW-152N intelligence-troop carrier.

The 1W18 or 1W19 command-staff vehicle, based on the BTR-60PB armored transport.

The BTR-60PBK command vehicle used by an artillery battery as an artillery observation and command vehicle.

A variety of command-staff vehicles based on the BTR-60PA and -60PB armored transports—from left to right, two R-145, one R-137, two R-145 and one NZ (B) MSR/PR.

Chapter 11: Special Armored Vehicles of the Soviet Army

The R-140BM command-staff vehicle (KW-radio set-channel former) (Russian: Radiostanziya KW R-140BM).

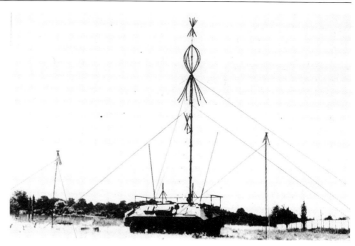

The R-145BM/L2 command-staff vehicle.

Vehicles of an intelligence unit, including two R-145BM command-staff vehicles.

The R-137B command-staff vehicle (KW radio set-channel former).

The R-145BM command-staff vehicle.

The PU-12 command vehicle of a battery command group.

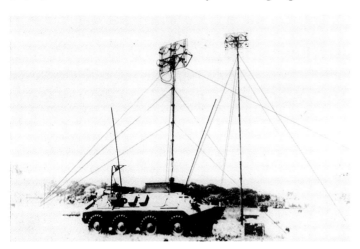
The R-409BM command-staff vehicle (directional radio set).

The PU-12 command vehicle of a troop AA unit, armed with ZSU-23-4M self-propelled AA guns.

P-238BT command-staff vehicle (teletype set).

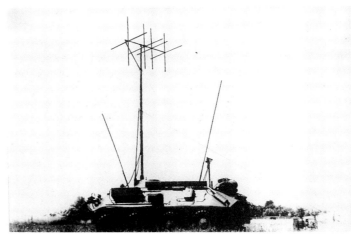
The P-240BT command-staff vehicle (telephone set) (Russian Telefonnaya apparatnaya P-240BT).

Chapter 11: Special Armored Vehicles of the Soviet Army

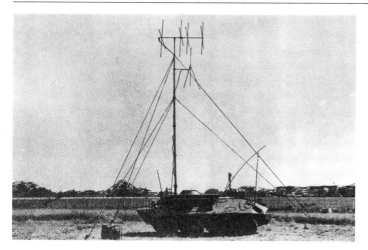

The P-241BT command-staff vehicle (teletype set).

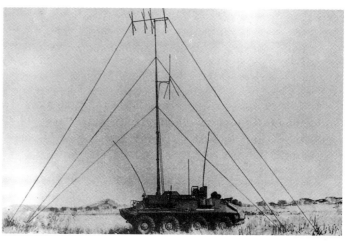

The intelligence center of the motorized rifle regiment/tank regiment IC (B) MRR/TR (Russian Uzel svayaz US (B) MSP/TP).

The PPO mobile post for training (Russian Podvishniy post obucheniya PPO), built for the students at the officers' college of the airborne troops, based on the BTR-50P armored transport: 1) had a removable body section with eleven training places for platoon leaders, 2) had a 12th seat, 3) had closeable hatches, 4) had observation devices, 5) had a seat for the training leader.

The PPO mobile post for training, based on the BMP-1 infantry combat vehicle, built for the students at the officer' college of the motorized rifle units in Ordshonikidze, North Caucasus.

The ZS-88 agitation and propaganda vehicle, based on the BTR-80 armored transport.

12

Recovery Tanks, Technical Service Vehicles and Tracked Barrier-crossing Technology

"The number of tanks needing to be towed grew from day to day. Out of the depths, means of salvage were conjured up. At the beginning of 1943, the 82nd Recovery company had already been equipped with heavy and medium towing tractors. The decision to rebuild twenty tanks as tractors did not come easy for our high command, for twenty tanks were almost a regiment. . . ." wrote the Russian General I. M. Golushko in his memoirs.[27] He began his war service as a technical officer in the 49th independent heavy tank regiment. In the first days of the German advance into Russia, the Red Army had to learn from painful experience that every immobilized tank became a "total loss" for lack of towing equipment. But when the Red Army was on the advance again, repairs in the field became the main source of replacements for losses. For example, in the 3rd Guard Tank Army during the operations at Lvov-Sandomierz, the number of repaired tanks and self-propelled guns exceeded the supply of combat vehicles that the army had had at the beginning of the operation. Several tanks were repaired and put back into service two or three times.[28] Still, in all, the policy in the first postwar years remained that only the hulls of mustered-out tanks were turned into recovery vehicles. As a rule, the armored chassis of the T-34, IS-2 and ISU-152 were so used.

Only in the mid-fifties, after extensive evaluative reports on the course of World War II were available and the economy had stabilized, was a start made in building specific recovery and repair vehicles that would be capable of use even under combat conditions. Tests showed that special armored vehicles are necessary to maintain the combat capability of, in particular, the armored troops. With an expectable loss of 10 to 12% of the combat technology, a unit would, toward the end of the fifth day of combat, only have 40 to 50% of its initial supply of usable tanks. But if the armored unit had appropriate towing and repairing equipment, then 60 to 70% of the disabled vehicles could be made ready for use again. This means that only after 12 to 16 days would the number be reduced to 40 to 50%. Now for the first time, concrete requests for the production of salvage tanks were made. They had to be capable of freeing stuck tanks and self-propelled guns, towing disabled vehicles to a repair unit, raising and salvaging armored vehicles damaged in combat, supplying technical service with appropriate special tools and spare parts, as well as being available for service in overcoming water barriers. It was further requested that these vehicles be usable as technical observation and command points, in order to radio for help or call up reinforcements. In the latter half of the fifties, the first BTS medium armored towing tractors were supplied to the tank regiments. In the ensuing times, special machines for recovery and technical service arose on the chassis of almost all armored vehicles. Their equipment itself was expanded more and more. Among others, the following devices were added: towing winches with cable drums, plow blades, underwater wading equipment with escape tubes, hydraulic presses, cutting and welding apparatus, special tools and folding jacks. In another change, the vehicles were equipped with tank or AA machine guns for self-defense. Sometimes even grenade launchers or recoilless guns were used.

To lift objects for field repairs, the recovery vehicles were finally equipped with various cranes. For especially great weights, every repair company received a crane tank. To remove obstacles, various tank hulls had a gripper installed and were introduced in engineer companies with the designation IMR. There were also armored vehicles for removing mines and creating lanes in minefields. Various tracked vehicles were used to create engineering machines for building positions or laying pipes for drinking water or for fuels and lubricants.

Along with the many versions of tracked and armored vehicles for technical service and engineering work, since the end of World

The BTS-2 medium armored towing tractor (1955 model).

---

27. Golushko, I. M., Panzer erwachen wieder, Berlin 1981, p. 150
28. Skatshko, P. G., Panzer und Panzertruppen, Berlin 1983, p. 356.

Chapter 12: Recovery Tanks, Technical Service Vehicles

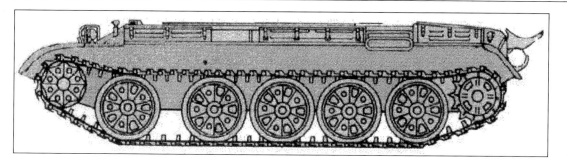

The BTS medium armored towing tractor (1954 model), called Recovery Tractor T-54T in the Warsaw Pact countries, consisted only of the armored hull of a T-54. The turret ring opening was covered with an armor plate including a hatch.

War II the most varried means of crossing have been developed. Among these are the bridgelaying tanks, which unlike those of all other European armies can only span gaps between 15 and 20 meters wide. To this day, only three different bridgelaying tanks have appeared. While the MTU or MTU-20 has been in service with the armored engineer units since the mid-fifties, the successor MTU-72 was designed only at the beginning of the eighties. Its introduction was limited to small numbers.

The main emphasis, though, was placed on floating transporters. The reason for this is the comparatively large number of rivers that are so wide that they could be spanned only by numerous segments attached together. Thus, the GSP-55, PMM and PMP ferries were originated to carry armored vehicles. For crossing water barriers with men, guns, cars or trucks the K-61, PTS-M and PTS-2 were developed.

Medium Armored Towing Tractor BTS-2
Bronetankoviy tyagatsha sredniy BTS-2 (obr. 1955 g.)

Made in: USSR, CSSR, Poland
Used in: USSR, East Germany, Finland, CSSR, Poland
Developed: 1955
Manufactured: 1955-1964/65
Crew: 2+1 men
Fighting weight: 32,000 kg
Length: 7050 mm
Width: 3275 mm
Height: 2500 mm
Armament: 1 7.62 mm SGMT or PKT tank machine gun (1000 rounds)
Powerplant: 12-cylinder W-2-55 Diesel engine (580 HP/427 kW)
Top speed: 50 kph
Range: 550 km

Medium Armored Towing Tractor BTS-3
Bronetankoviy tyagatsha sredniy BTS-3 (obr. 1963 g.)

Made in: USSR
Used in: USSR/CIS/Russia, East Germany, Poland, CSSR
Developed: 1962-63
Manufactured: 1963-1981
Crew: 2+1 men
Fighting weight: 42,000 kg
Overall length: 9740 mm (without dozer 9105 mm)
Hull length: 7120 mm
Width: 3385 mm
Height: 2650 mm
Armament: 1 7.62 mm SGMT or PKT tank machine gun (1000 rounds)
Powerplant: 12-cylinder W-2-55 or W-2-55A Diesel engine (580 HP/427 kW, later 620 HP/456 kW, finally 680 HP/500 kW)
Top speed: 45 kph
Range: 270 km

The BTS-2 medium armored towing tractor (1955 model), seen during the Warsaw Pact "Weapon Brotherhood 70" maneuvers, October 12-18, 1970.

The BTS-3 medium armored towing tractor (1963 model) of the 2nd Shock Army, formerly stationed in East Germany, seen in the streets of Dresden.

Medium Armored Towing Tractor BTS-4
Bronetankoviy tyagatsha sredniy BTS-4 (obr. 1964 g.)

Made in: USSR, CSSR
Used in: USSR/CIS/Russia, CSSR, East Germany, Finland
Developed: 1963-64
Manufactured: 1964-1982
Crew: 2+1 men
Fighting weight: 35,000 kg
Length: 7120 mm
Width: 3275 mm
Height: 2850 mm
Armament: 1 7.62 mm SGMT or PKT tank machine gun (1000 rounds)
Powerplant: 12-cylinder W-2-55 or W-2-55A Diesel engine (580 HP/427 kW), later 620 HP/456 kW, finally 680 HP/500 kW)
Top speed: 50 kph
Range: 400 km

Armored Recovery and Repair Vehicle BREM-1
Bronirovanaya remontno-evakuatsionaya maschina BREM-1

Made in: USSR/CIS/Russia
Used in: USSR/CIS/Russia
Developed: 1978-79
Manufactured: since 1981-82
Crew: 3 men
Fighting weight: 42,000 kg
Length: 7100 mm
Width: 3460 mm
Height: 2400 mm
Armament: 1 12.7 mm NSWT heavy machine gun (500 rounds)
Powerplant: 12-cylinder (multifuel) W-46 Diesel engine (780 HP/573 kW), W-46-6 (792 HP/582 kW), W-84 (840 HP/618 kW), W-84A (842 HP/619 kW)
Top speed: 60 kph
Range: 480 km

The BTS-3 medium armored towing tractor as a T-55TK crane tank of the East German Army.

The crane could lift fifteen tons. The BRDM armored reconnaissance and scouting vehicle, weighing 5.6 tons, was no great challenge for this crane.

Chapter 12: Recovery Tanks, Technical Service Vehicles

On the other hand, the BMP-1 infantry combat vehicle of Motorized Rifle Regiment 27, weighing 13.5 tons, tested the crane's limits during the "Running Gear 87" drill.

The BTS-4 medium armored towing tractor (1964 model) was known in the East German Army as T-55T.

Armored Recovery and Repair Vehicle BREM-2
Bronirovanaya remontno-evakuatsionaya maschina BREM-2 (obr. 1982 g.)

Made in: USSR/CIS/Russia, CSSR
Used in: USSR/CIS/Russia, CSSR, East Germany
Developed: 1982
Manufactured: since 1986
Crew: 3 men
Fighting weight: 13,600 kg
Length: 6577 mm
Width: 3184 mm
Height: 2280 mm
Armament: 1 7.62 mm PKT or PKT-M tank machine gun (1000 rounds)
Powerplant: 6-cylinder UTD-20 Diesel engine (295 HP/217 kW)
Top speed: 65 kph (road), 7 kph (water)
Range: 600 km

Armored Recovery and Repair Vehicle BREM
Bronirovana remontno-evakuatsionaya maschina BREM (obr. 1988 g.) (na base BTR-80)

Made in: CIS/Russia
Used in: CIS/Russia
Developed: 1988
Manufactured: since 1988
Crew: 4 men
Fighting weight: 14,000 kg
Length: 7850 mm
Width: 2900 mm
Height: 2833 mm
Armament: 1 7.62 mm PKT or PKT-M tank machine gun (1000 rounds)
Powerplant: 8-cylinder (multifuel) Diesel engine (260 HP/919 kW)
Top speed: 90 kph (road), 9 kph (water)
Range: 600 km

Rear view of a BTS-4 medium armored towing tractor (1964 model). Two 200-liter fuel tanks could be attached to the spur.

The BTS-4 medium armored towing tractor with an added underwater air tube.

Major overhauling of BTS-4 medium armored towing tractors (1964 model).

Technical Service Vehicle MTP
Maschina technikoi pomosch MTP (na base MT-LB)

Made in: USSR/CIS/Russia, Bulgaria
Used in: USSR/CIS/Russia, Bulgaria, East Germany
Developed: 1979-80
Manufactured: since 1980
Crew: 3 men
Fighting weight: 12,300 kg
Length: 6800 mm
Width: 2850 mm
Height: 2300 mm
Powerplant: 8-cylinder JaMZ-238W Diesel engine (240 HP/177 kW)
Top speed: 55 kph (road), 6.5 kph (water)
Range: 400 km

The BTS medium armored tractor, based on the SU-122-54 self-propelled artillery mount (1949 model), seen in Moscow on November 7, 1977. The NATO name was Recovery Tractor M.1977, but the chassis was often called that of the T-62-based self-propelled mount, because of its changed intervals between the road wheels.

The BTS-4 medium armored towing tractor during training.

The BREM-1 armored repair and recovery vehicle, built on the basis of the T-72 medium tank.

Chapter 12: Recovery Tanks, Technical Service Vehicles

Bridgelaying Tank MTU
Mostoukladshik tankovy ustrostvo MTU

Made in: USSR
Used in: USSR, Afghanistan, Bulgaria, East Germany, Egypt, Finland, Israel, Poland, Syria
Developed: 1957
Manufactured: 1957-1968
Crew: 2 men
Fighting weight: 34.0 tons
Overall length: 12,300 mm
Hull length: 7200 mm
Width: 3270 mm
Height: 3400 mm
Armament: 1 12.7 mm DSchK-M heavy machine gun (500 rounds)
Powerplant: 12-cylinder W-2-54 Diesel engine (520 HP/382 kW) or W-2-55 (580 HP/427 kW)
Top speed: 50 kph
Range: 450 km

At the end of 1998 the tank factory in Chelyabinsk presented the BREM-80U armored repair and recovery vehicle on the basis of the T-80U medium tank. The chassie corresponds to the T-80U medium tank, while the rest corresponds to the BREM-1 armored repair and recovery vehicle on the basis of the T-72 medium tank.

The BREM-2 armored repair and recovery vehicle (1982 model), made since 1986 on the basis of the BMP infantry combat vehicle.

The BREM armored repair and recivery vehicle, based on the BMP-3.

Above and Below: **The MTP-1 technical service vehicle, produced on the basis of the BTR-50PK armored transporter.**

1. UF tube; 2. Spare parts; 3. Transport rack, 1706 x 1400 mm; 4. 12.7 mm NSWT AA machine gun; 5. 19-ton crane; 6. Plow boade; 7. Steering wheel; 8. rubber-link track section; 9. Road wheel; 10. Drive wheel; 11. Load [kaken]; 12. Winch; 13. Crane boom; 14. Hydraulic cylinder; 15. 200-liter fuel barrels.

The MTP-LB technical service vehicle, based on the MT-LB multipurpose towing and transport vehicle.

The IMR engineer vehicle for removing obstacles (Russian Inshenernaya maschina rasgrasheniya), built on the basis of the T-55 medium tank.

The IMR-2 engineer vehicle for removing obstacles, based on the T-72 medium tank.

Chapter 12: Recovery Tanks, Technical Service Vehicles

The GMZ track-driven minelayer (Russian Gusenichnii minno-zagratidel), produced on the basis of the 2-S3 "Acacia" self-propelled artillery mount.

The MTK mine removal tank with UR-67 firing apparatus, for removing mines with [sharp bend] or antagonistic ignition in a lane up to 15 meters wide. The vehicle was built on the BTR-50PK armored transporter.

The MTK-2 mine removal tank with UR-77 firing apparatus. The chassis of the 2-S1 "Carnation" self-propelled artillery mount was used to build this tank.

On the basis of the SU-122-54 self-propelled artillery mount or the BTS medium armored towing tractor, a mine removal tank with KMT-5 (roller section) was built for use in Afghanistan.

The BTS-4 medium armored towing tractor was also fitted with the KMT-5 mine removal device for use in Afghanistan.

The BTU-80 armored dozer (Russian Buldozernoy tankovoy ystroistvo), based on the T-80 medium tank.

The MTU bridgelaying tank (1957 model) with 12.3-meter vehicle bridge, capable of carrying 50 tons, seen in June 1965.

The MTU bridgelaying tank (1957 model), seen during drills by the Soviet military group in Germany and the East German People's Army on June 15, 1967.

Extending and placing the 12.3-meter vehicle bridge.

The MTU-20 bridgelaying tank (1967 model) with 20-meter vehicle bridge, its carrying capacity stated as 60 tons.

The GSP-55 self-propelled ferry consisted of two half-ferries that were linked in the water. The picture shows the right half-ferry.

The MTU-72 bridgelaying tank, using the hull of the T-72B tank.

Chapter 12: Recovery Tanks, Technical Service Vehicles

The capacity of the GSP-55 self-propelled ferry was 52 tons.

After the two half-ferries were linked in the water, the pontoons were removed.

The PMM-2 amphibious ferrying vehicle was built on the chassis of the MT-T heavy multipurpose towing and transport vehicle.

Two half-ferries with pontoons in place.

In the water, the vehicle is driven by two screw propellers mounted on the rear end; the left one can be seen clearly in this photo from October 1988.

139

Separating the two floating segments.

Although the PMM-2 amphibious ferrying vehicle no longer consists of two half-ferries, its capacity is sufficient to cross water barriers with a T-64 medium tank (fighting weight 38 tons).

To ferry men, weapons and equipment, the PTS medium amphibious transporter (Plavayushii transporter, srednii) was introduced into the Soviet Army at the end of the fifties.

The capacity of the vehicle was over five tons. This PTS is crossing a water obstacle with a ZIL-151 truck.

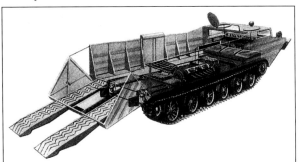

In the Warsaw Pact forces, this transporter was called Swim Wagon K-61. It is seen laden with a BTR-40 armored transporter (fighting weight 5.3 tons) on April 22, 1961.

Chapter 12: Recovery Tanks, Technical Service Vehicles

The PTS-M improved medium amphibious transporter had a capacity of ten tons.

The PTS-M transporter could carry up to seventy soldiers.

Since the capacity could be increased again and the rear of the PTS-M was exposed when heavy weights were loaded, the PTS-2 has a support apparatus.

The PTS-2 medium amphibious transporter was introduced into the Soviet Army as of 1985. As with the PMM-2, this vehicle also used the chassis of the MT-T heavy multipurpose towing and transport vehicle.

13

Special Towing Tractors on Tracked Chassis

Building on the experience of World War II and the requests for full motorization of the ground combat troops, the Soviet armament industry developed various tracked towing tractors between 1948 and 1953. Depending on their towing power, they were divided into light (Artilleriskii tyagatsha-legkii, AT-L), medium - srednii, AT-S) and heavy artillery tractors (-tyashalii, AT-T).

For service under the extreme conditions of the Far North and the roadless territory beyond the Urals, the GT-S (Gusenitshii traktor-srednii) medium towing tractor was developed. This vehicle was also built with a somewhat larger rear bed and a more powerful engine than the GT-T (tyashelii). Through its extremely low ground pressure, it can be used very well in the Antarctic and the tundra.

Along with the use of tracked tractors as means of towing and transport, the ATS-59 medium artillery tractor and the GT-T were also used as cabs for semi-trailers transporting anti-aircraft rockets in desolate regions.

The AT-P artillery tractor and the MT-L multipurpose transporter and tractor were introduced as universally useful light armored tractors for the artillery. With their self-supporting closed hulls, they can protect their crews from the effects of light enemy weapons and shell fragments. With the possibility of sealing the hull hermetically, these vehicles can also pass through terrain that has been contaminated with means of mass destruction. The improved MT-LB version (mnogozelevogo gusenitshnogo transportera-tyagatsha-legkii) was introduced into the Soviet Army toward the end of the sixties. Since then it has been used not only as a tractor and means of transport, but also as a tank destroyer, anti-aircraft, reconnaissance, repair shop, artillery observation and command vehicle. The MT-LB is built in three sections. In its bow is the gearbox, in front of the driver and commander; in the middle the eight-cylinder Diesel engine; and in the rear the cargo space. The weight of the load that can be carried is 2,500 kilograms. The maximum allowable trailer weight is 6,500 kilograms. Thirteen seats are available in all. The MT-LB is armed with a 7.62 mm tank machine gun. While underway, it is possible for two soldiers to fire hand weapons at the enemy through close-combat loopholes. For entrance and exit there are hatches in the roof and two large ones at the rear. The vehicle is amphibious and is propelled by the tracks in water. It also has rudders, not to steer with, but to direct the water and keep it close to the tracks so that the vehicle could move in the wake. In addition, the deflector attached to the front can be folded upward. The top speed is 61.5 kph on the road and 6 kph in the water. The special equipment also includes the radio station, the speaker system, targeting and observation devices, the driver's night-vision device, a warning system, heating for the fighting compartment, and protection from nuclear weapons. For use in the Far North, wider tracks are installed. Since 1986 the rubber-and-metal track links used on the 1W13 and 1W16 command vehicles have begun to be used here as well.

The AT-L light artillery tractor (1948 model) could tow vehicles up to 6,000 kilograms; its allowable load limit was 2,000 kilograms.

Chapter 13: Special Towing Tractors on Tracked Chassis

Heavy Artillery Tractor AT-T
Artilleriskaya tyagatsha-tyashely AT-T (obr. 1950 g.)

Made in: USSR
Used in: USSR, East Germany
Developed: 1950
Manufactured: 1950-1962
Crew: 3+14 men
Fighting weight: 25,000 kg
Length: 6990 mm
Width: 3170 mm
Height: 2845 mm
Powerplant: 12-cylinder W-401 Diesel engine (410 HP/302 kW)
Top speed: 35.5 kph
Range: 750 km

Artillery Towing Tractor of the Infantry AT-P
Artilleriskaya tyagatcha-pechoti AT-P (obr. 1954 g.)

Made in: USSR
Used in: USSR, CSSR
Developed: 1953
Manufactured: 1954-1970
Crew: 3 men
Fighting weight: 7200 kg
Length: 4500 mm
Width: 2500 mm
Height: 1800 mm
Armament: 1 7.62 mm SGMT or PKT tank machine gun (1000 rounds)
Powerplant: 6-cylinder ZIL-123F Diesel engine (110 HP/81 kW)
Top speed: 50 kph
Range: 350 km

Medium Towing Tractor GT-S
Gusenitshii traktor-sredniy GT-S (obr. 1953 g.)

Made in: USSR
Used in: USSR, East Germany
Developed: 1952
Manufactured: 1953 to 1958-59
Crew: 2+6 men
Fighting weight: 4650 kg
Length: 5230 mm
Width: 2560 mm
Height: 1910 mm
Powerplant: 4-cylinder GAZ-47 gasoline engine (85 HP/63 kW)
Top speed: 35 kph
Range: 450 km

Heavy Towing Tractor GT-T
Gusenitshii traktor-tyashelyi (obr. 1950 g.)

Made in: USSR
Used in: USSR
Developed: 1949-50
Manufactured 1950-1966
Crew: 3+10 men
Fighting weight: 10,200 kg
Length: 7600 mm
Width: 3140 mm
Height: 2200 mm
Powerplant: 6-cylinder 13-6 Diesel engine (200 HP/147 kW)
Top speed: 45 kph (road), 8 kph (water)
Range: 250 km

Multipurpose Towing and Transport Vehicle MT-LB
Mnogozelevogo gusenitshnogo transportera-tyagatcha-legkiy (variant B) (obr. 1970 g.)

Made in: USSR/CIS/Russia, Bulgaria
Used in: USSR/CIS/Russia, Bulgaria, CSSR, East Germany, Finland, Hungary, Iraq, Poland, Vietnam, Yugoslavia
Developed: 1969
Manufactured: 1970-1977 (1st series), 1977-1980 (2nd), 1980-1986 (3rd), since 1986 (4th)
Crew: 2+8 to 11 men
Fighting weight: 11,900 kg
Length: 6454 mm
Width: 2850 mm
Height: 1865 mm
Armament: 1 7.62 mm PKT or PKT-M tank machine gun (1000 rounds)
Powerplant: 8-cylinder JaMZ-238 or JaMZ-238W Diesel engine (240 HP/177 kW)
Top speed: 61.5 kph (road), 6 kph (water)
Range: 475 km

The AT-L light artillery tractor was used mainly by the heavy grenade-launcher units of the Soviet Army.

In the tank-destroyer units of the motorized rifle divisions, the AT-P infantry artillery tractor (1954 model) was used to tow antitank guns.

The AT-P infantry artillery tractor with a T-12 100 mm antitank gun, seen during the Warsaw Pact "October Storm" maneuvers in 1965.

The AT-S medium artillery tractor (1956 model) was suitable for towing up to 14,000 kilograms. Its allowable load limit was stated as 3,000 kilograms.

The AT-S towing a 152 mm cannon howitzer (1937 model).

The AT-S medium artillery tractor with a dozer blade as the OST bulldozer.

The ATS-59 medium artillery tractor had fourteen seats in the rear, a pulling power of 13,025 kilograms, and a load limit of 3,000 kilograms.

Chapter 13: Special Towing Tractors on Tracked Chassis

The ATS-59G artillery tractor as a troop carrier for roadless areas.

The AT-T heavy artillery tractor (1950 model) was intended to pull loads up to 25,000 kilograms. Up to 5,000 kilograms of heavy material could be carried on its rear bed.

The running gear of the AT-T tractor was almost identical to the T-54 medium tank, and the 12-cylinder W-401 Diesel engine was a slightly modified W-2-54 tank engine.

Equipped with a plow blade, the vehicle became the BAT dozer (based on the AT-T 401).

The BAT-M dozer (based on the AT-T 405 mu) was made especially for position and road building. Instead of a rear bed, this vehicle has a loading crane.

For position building, the MDK-2M excavator (based on the AT-T 409 mu) was built (Russian Kotlovannaya maschina).

For digging trenches, the BTM excavator (based on the AT-T 409) and BTM-3 (based on the AT-T 409 u) were built. (Russian Bistrochodnaya transcheynaya maschina)

For work under especially rough climatic and terrain conditions, the GT-S medium towing tractor (1953 model) was built.

With an allowable trailer weight of 5,000 kilograms and a load limit of 1,000 kg, the vehicle was essentially a light towing tractor.

The GT-T heavy towing tractor (1950 model) was amphibious, could tow up to 10,000 kilograms, and had a load limit of 4,000 kg.

There were up to fourteen seats in the back of the GT-T, thus the vehicle was sometimes used as an armored troop carrier.

Chapter 13: Special Towing Tractors on Tracked Chassis

As a towing tractor for artillery weapons, transporter for their crews, ammunition and equipment, the MT-LB multipurpose towing and transport vehicle was introduced. The vehicle's load limit was stated as 2,000 kilograms for a trailer and 2,500 kg without. The allowable trailer weight was some 6,500 kg, with the towing power at the hook 7,270 kg without and 8,790 kg with a load. The picture shows the MT-LB as an artillery tractor for the 2450-kilogram 122 mm howitzer (1938 model).

Winter driving tests with an MT-LB multipurpose towing and transport vehicle. In regions with extreme climates, the vehicle was also used as an armored transport for motorized riflemen.

The MT-LB was an amphibious transport vehicle; water deflectors were attached to its forward track aprons for water travel.

For transporting men, freight, weapons and equipment, as well as for use as a towing tractor under very rough climatic and terrain conditions, the Soviet Army introduced two-section transporters (DT = dvuchrasdelnyi transporter) with various load limits. This is a DT-20 two-section transporter, which could carry loads up to a total of 20,000 kilograms.

Shift lever and gearbox, part of the instrument panel, power source for the radio, driver's periscope, brake and gas pedals of the MT-LB.

The V8-cylinder JaMZ-238W Diesel engine produced 240 HP (177 kW) and was situated behind the driver in the left center of the MT-LB.

For loads up to 30,000 kilograms, the two-section DT-30 was built. The allowable gross weight was stated as 59 tons. The top speed was 37 kph. With a track width of 1100 mm, the ground pressure was 0.27 kg/sq.cm.

Chapter 13: Special Towing Tractors on Tracked Chassis

On the chassis of the MT-T, the BAT-2 road-building machine was built.

Numerous components of the T-64 chassis were used to build the MT-T heavy multipurpose towing and transport vehicle.

For the MDK-3 excavator, once again the MT-T heavy multipurpose towing and transport vehicle provided the basis.

The BTM-4 excavator was also built on the MT-T chassis.

14

Armored Vehicles of Air-landed Units

In June 1946 the airborne troops were separated from the ranks of the air combat forces and subordinated directly to the Ministry of Defense. This was linked with the formation of a command for the airborne troops, directed by the commander of the airborne troops. Using the rifle divisions as a model, the paratroop and airborne units were set up. Parallel to the organizational changes, the airborne troops were now supplied with new vehicles and equipment. In addition, the numbers of automatic-fire weapons, artillery guns, grenade launchers, and antitank and anti-aircraft guns were increased considerably. For the first time, the airborne units were supplied with armored vehicles. In 1947 the K-73 self-propelled gun and K-75 armored troop carrier appeared, followed in 1949 by the ASU-76 self-propelled gun, in 1950 by the K-78 armored troop carrier, in 1951 by the ASU-57 airborne self-peopelled gun, in 1954 by the ASU-57P and in 1956 the ASU-85. To land heavy weapons, light self-propelled guns and small vehicles, appropriate freight parachutes and gliders were developed. The ASU-57 proved to be very mobile, including in rough terrain. This was attained through the very meager ground pressure of 0.35 kp/sq.cm. with a gross weight of 3,300 kilograms. The four-stroke, 50 HP gasoline engine gave the vehicle a top speed of 45 kph. The installed 57 mm antitank gun had a rate of fire of six to ten shots per minute. The ammunition included shells with armor-piercing tracer, explosive splinter and fog cartridges. The effective range of the explosive splinter shells was 6,000 meters, that of the armor-piercing shells 1,250 meters. As of 1956 the introduction of the ASU-85 began. As a self-propelled gun mount with a gross weight of 15 tons, it could not be dropped by parachutes. It had to be brought to the enemy's hinterlands in transport planes to increase the airbore troops' supply of armored vehicles. Because of its small height, it could be moved across country quite inconspicuously, was easy to camouflage, and offered little target area. The crew consisted of three men. The commander was simultaneously aiming gunner and radioman. The others were the driver and loader.

At the beginning of the seventies, airborne units were first assigned to the ground combat forces as well. While the airborne divisions were directly subordinate to the military high command, the airborne assault brigades were used by the operative leadership. "With the advent of nuclear rocket weapons, the role of the airborne troops, which can utilize the results of atomic attacks quickly, increased," explained the deputy defense minister of the USSR, Army General S. Sokolov.[29] The Soviet command saw it as an absolute necessity to arm the airborne units more strongly and in more variety, so as to fulfill their new tasks and combat requirements. As a result, the airborne units were slowly supplied with the combat vehicles of the BMD (Boevaya maschina desanta) landing troop in place of the ASU-57 and ASU-85 self-propelled guns. Officially, the BMD were first shown to the Warsaw Pact states in the 1970 "Dvina" maneuvers and landed from the air by transport planes. These vehicles were fitted with the same turret as the infantry BMP combat vehicles. The turrets included a 73 mm 2A28 tank gun, a 7.62 mm PKT tank machine gun, and a launcher for 9M14 "Malyutka" antitank rockets. In addition, the BMD was armed with two tank machine guns at the right and left sides of the bow. The aiming periscope, consisting of one day and one electrooptic night-sight device, serves both for spotting targets and aiming the weapons, including the two machine guns in the bow. With the tank gun, targets at ranges up to 1,600 meters can be fired on, and the range of the antitank rockets is between 500 and 3,000 meters. The crew consists of the commander, the aiming gunner, the driver and four paratroops. The closed body of the BMD enables it to operate in contaminated areas.

Meanwhile, the most varied variations have been developed. On the basis of the BMD, the BTR-D armored transporter was introduced in 1974, the BMD-2 landing-troop combat vehicle with 2A42 30 mm gun in its turret in 1983, an armored BREM-D recovery and repair vehicle in 1984, and several command and observation vehicles since 1985. Among these are the BMD-1K, BMD-1PK, BMD-1R, BMD-1KSh, BMD-2K, R-440ODB and 1W119.

At the 1985 May Day Parade in Moscow, the Soviet government introduced the 2-S9 "Anona" self-propelled gun. With a 120 mm tank gun, this became a high-firepower support vehicle for the airborne troops. "The gun can fire directly, out of an open firing position, like a cannon. From a covered firing position it can fire on its target like a howitzer, from low spots and gullies it can fire like a classic mortar."[30] The vehicle, designated as a 120 mm gun on a

---

29. Sobik, E., Die sowjetischen Luftlandetruppen, Frankfurt am Main 1976, p. 510.
30. Knyaskov, R. W., Vojenniye Snaniya No. 05, Moscow 1988, p. 8.

Chapter 14: Armored Vehicles of Air-landed Units

One of the first vehicles of the paratroops was the K-73 self-propelled mount with the 57 mm gun (model 1949). The gross weight was 3.40 tons, the length 5366 mm, the width 2085 mm and the height 1400 mm. The top speed was 54 kph on the road and 7.8 kph in water. The range was 265 km. It was driven by a 70 HP 51.5 kW gasoline engine.

The first vehicle supplied to the paratroops in large numbers was the ASU-57 (1951 model) airborne self-propelled mount. This one was photographed in September 1963.

self-propelled mount, is amphibian and can be dropped from the air by parachute.

At the end of the eighties, the successor model BMD-3 went into production. Its armament equals that of the BMD-2. The hull, though, was newly designed. Above all, the use of the running gear derived from the PT-76 suggests a considerable weight gain for the vehicle.

Airborne Self-propelled Gun ASU-57
Aviadesantnaya samochodnaya ustanovka ASU-57 (obr. 1951 g.)

Made in: USSR
Used in: USSR, Egypt, Vietnam, Yugoslavia
Developed: 1950
Manufactured: 1951-1962
Crew: 3 men
Fighting weight: 3300 kg
Overall length: 5750 mm
Hull length: 3350 mm
Width: 2086 mm
Height: 1460 mm
Armament: 1 57 mm TSh-51 or TSh-51M L/73 antitank gun (30 rounds)
Powerplant: 4-cylinder B-70, A-72 or M-20E gasoline engine (50 HP/37 kW, later 55 HP/40.5 kW)
Top speed: 45 kph
Range: 250 km

Airborne Self-propelled Gun ASU-85
Aviadesantnaya samochodnaya ustanovka ASU-85 (obr. 1956 g.)

Made in: USSR
Used in: USSR, Poland
Developed: 1955
Manufactured: 1956-1969
Crew: 4 men
Fighting weight: 15,000 kg
Overall length: 7730 mm
Hull length: 5460 mm
Width: 2950 mm
Height: 2250 mm
Armament: 1 85 mm D-70 L/53 gun (45 rounds)
1 7.62 mm SGMT or PKT tank machine gun (2000 rounds)
Powerplant: 6-cylinder W-6R Diesel engine (210 HP/155 kW)
Top speed: 45 kph
Range: 360 km

The crew of the ASU-57 numbered three men, but there was space in the vehicle for two more paratroopers.

Attack by a paratroop unit, equipped with ASU-57 airborne self-propelled gun mounts.

To equip the paratroops with more powerful armor-piercing weapons, the ASU-85 airborne self-propelled mount (1956 model) was introduced.

The ASU-57 airborne self-propelled mount (1951 model) (1951 model), seen in Czechoslovakia during Warsaw Pact "Moldau" maneuvers in September 1966.

The ASU-85 airborne self-propelled mount could not be dropped with parachutes. It had to be brought to the enemy's hinterlands in transport aircraft.

Chapter 14: Armored Vehicles of Air-landed Units

Paratroops land from the sea with an ASU-85 airborne self-propelled mount.

Since the early sixties, the ASU-85 airborne self-propelled mount has gradually been armed with a 12.7 mm DSchKM AA machine gun.

Combat Vehicle of the Airborne Troops BMD-1 and BMD-2

Boewaya maschina desanta BMD-1 (obr. 1968 g.), BMD-2 (obr. 1983 g.)

Made in: USSR/CIS/Russia
Used in: USSR/CIS/Russia
Developed: 1968 (BMD-1), 1983 (BMD-2)
Manufactured: 1968-1985 (BMD-1), 1983 to 1995-96 (BMD-2)
Crew: 3+4 men

	BMD-1	BMD-2
Fighting weight:	7200 kg	8000 kg
Overall length:	5400 mm	5970 mm
Hull length:	5400 mm	5400 mm
Width:	2630 mm	2700 mm
Height:	1970 mm	2180 mm
Height, lowered:	1620 mm	1830 mm
Armament:	1 73 mm 2A28 gun (40 rounds)	1 30 mm 2A42 gun (300 rounds)
	1 7.62 mm PKT or PKT-M machine gun (2000 rounds)	
	2 7.62 mm AKM-47 Mpists	2 5.45 mm AKM-74 Mpistols
	3 PALR 9M14 or 9M14M "Malyutka"	3 PALR 9M111 "Fagot" or 9M113 "Konkurs"

Powerplant: 6-cylinder 5D20 Diesel engine (240 HP/176 kW)
Top speed: 60 kph (road), 9 kph (water)
Range: 500 km

A BMD-1 is being dropped from an airplane.

With the introduction of the BMD-1 landing-troop combat vehicle (1968 model), the paratroops obtained a relatively heavily-armed vehicle that could be dropped from the air.

In order to decrease the impact on the ground, braking rockets were applied before landing.

153

To prevent torsion bars being broken by the impact with the ground, the chassis was lowered before being dropped.

An air-landed unit on the move with BMD-1s.

A BMD-1 landing-troop combat vehicle at a driving school.

At the beginning of the seventies, a brace for attaching the 9P135 or 9P135M launcher was attached to the roof of the turret. Thus, the BMD-1P landing-troop combat vehicle became capable of launching the 9M111 "Bassoon" (NATO code: AT-4/SPIGOT) and 9M113 "Competition" (Russian "Konkurs"; NATO code: AT-5/SPANDREL) anti-tank rockets. As of this production run, the vehicles were also given new-type road wheels.

Chapter 14: Armored Vehicles of Air-landed Units

Front view of a BMD-1P, with the cover of the left bow machine gun easy to see.

Rear view of a BMD-1P; the vehicle was driven in the water by two water jets.

A BMD-1P during training.

The BMD-1P landing-troop combat vehicle with built-on 9P135 launcher.

The BMD-2 version built since the beginning of the eighties (1983 model) was armed with the 30 mm 2A28 machine gun. With an elevation to 74 degrees, this weapon can also be used against low-flying air targets.

During the armed clashes in Georgia in April 1989, the BMD-2 was seen publicly for the first time.

155

The newest vehicle of the Russian paratroops is the BMD-3. On a completely new hull, a turret very much like that of the BMP-2 infantry combat vehicle was mounted.

The BTR-D armored transporter and BMD-1P landing-troop combat vehicle are seen during training of an air assault regiment then stationed in East Germany.

With the BTR-D armored transporter (1974 model), up to ten paratroops can be carried in addition to the three crewmen.

For artillery support of paratroops, the 2-S9 "Bottletree" self-propelled mount (Russian "Anona") was introduced.

On the basis of the BMD landing-troop combat vehicle, numerous command and observation vehicles were created. This is the BMD-1KSch command-staff vehicle, pulling a ZU-23 23 mm twin AA gun.

Chapter 14: Armored Vehicles of Air-landed Units

The arc of elevation of the 120 mm gun extends from -4 to +80 degrees. Thus the gun can be used for both direct and indirect fire. Although the gun is housed in a turret, the traverse is said to cover only 35 degrees to each side. The rate of fire is 6 to 8 shots per minute, the maximum range is 8800 meters.

Rear view of a BMD-1KSch command-staff vehicle.

At the IDEX 1993 in Abu Dhabi, the newest self-propelled gun mount for landing troops was displayed. The 2-S31 "Vein" (Russian "Vena") self-propelled mount was built on the BMD-3 chassis and is armed with a long 120 mm gun.

Schematics Gallery

T-34-85 Medium Tank (1947 model) The Driver's Compartment:
1. Driver's seat; 2. Steering levers; 3. Gas pedal; 4. Brake pedal; 5. Clutch pedal; 6. Compressed air bottles; 7. Dashboard light; 8. Dashboard; 9. Periscope; 10. Springs; 11. Tachometer; 12. Speedometer; 13. Speaker set no. 3 for driver; 14. Starter button; 15. Roof hatch latch; 16. Signal button; 17. Front suspension shaft; 18. Regulator control; 19. Shift lever; 20. Electric box with controls for electric devices.

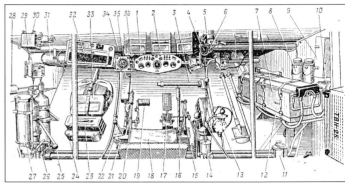

PT-76B light amphibian tank (1958 model) crew compartment.

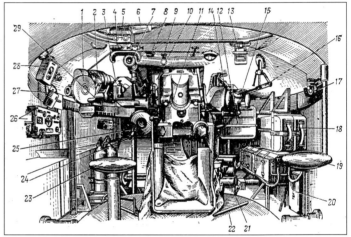

PT-76B light amphibian tank (1958 model) crew compartment.

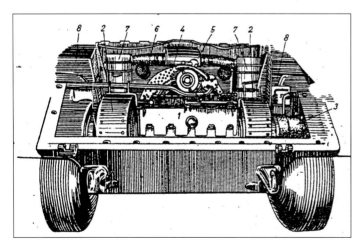

SU-100M medium self-propelled gun (1947 model) The Transmission Area: 1. Gearbox; 2. Clutches with brakes; 3. Side driveshafts; 4. Main clutch with ventilator; 5. Air intake; 6. Ducting; 7. Air filter; 8. Fuel tank.

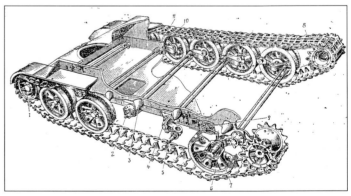

T-54 and T-55 Medium tanks, drive train.

Schematics Gallery

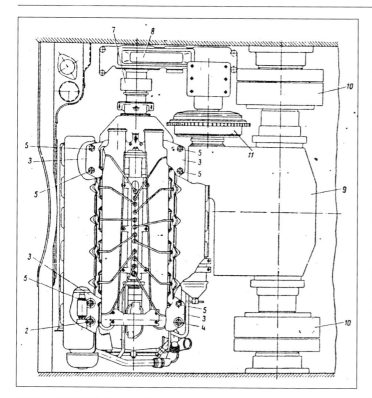

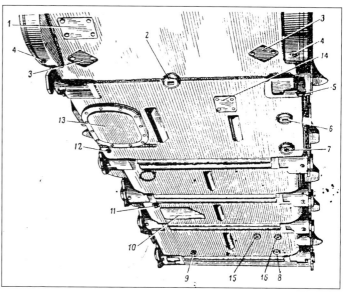

T-54 and T-55 Medium tanks.

T-54 and T-55 Medium tanks. Engine Layout: 2, 4, 5. Screws; 3. Engine brackets; 7. Engine-gearbox link; 8. Intermediate gears; 9. Gearbox; 10. Steering gears; 11. Main clutch.

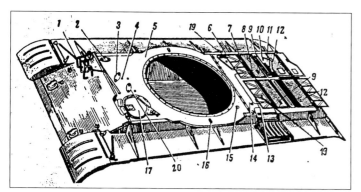

T-54 and T-55 Medium tanks, upper hull.

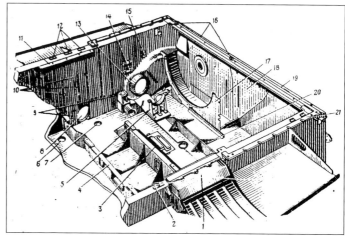

T-55A Engine Room: 1. Armor plate; 2. Attachment; 3. Front gearbox attachment; 4. Engine block frame; 5. Rear gearbox attachment; 6. Fuel line opening; 7. Opening for oil line from gearbox; 8. Opening for sway damper; 9. Intermediate gear attachments; 10. Screw couplings; 11. Attachment bracket; 12. Brace; 13. Bracket for driveshafts; 14. Brake attachment screw; 15. Brake spring attachment; 16. Lever for hatch over air filter; 17, 18. Air filter attachment panel; 19. Upper rear plate; 20. Reinforcing rib; 21. Attachment bar.

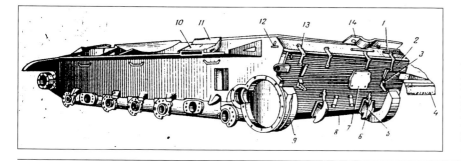

Hull of the T 55A Tank, rear view: 1. Upper rear plate; 2. Central rear plate; 3. Bracket for towing block; 4. Folding rear shield; 5. Towing hook; 6. Towing hook lock; 7. Hatch cover; 8. Lower rear plate; 9. Driveshaft housing; 10. Exhaust bracket; 11. Armor plate; 12. Taillight bracket; 13. Road wheel bracket; 14. Shield over ventilator.

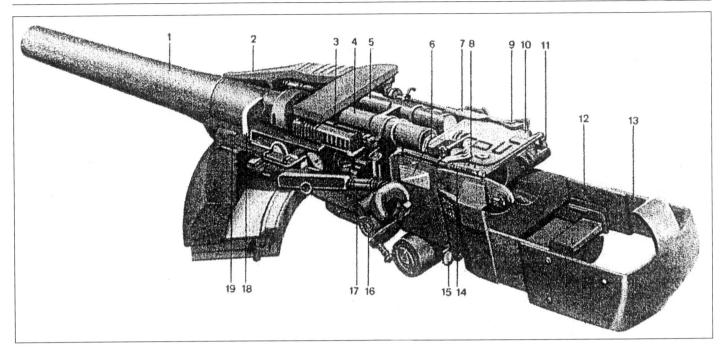

100 mm D-10T2S L/56 tank gun (T-54M and T-55 medium tanks): 1. Barrel; 2. Shield; 3. Oil reservoir; 4. Recoil brake; 5. Pneumatic recuperator; 6. Hydraulic cylinders for recoil brake; 7. Machine gun; 8. Automatic lock; 9. Tension lever; 10. Breech with lock; 11. Breech handle; 12. Immobile part of frame; 13. Folding part of frame; 14. Breech-locking device; 15. Manual trigger; 16. Elevation control; 17. Targeting scope; 18. Gun cradle; 19. Turret armor frame.

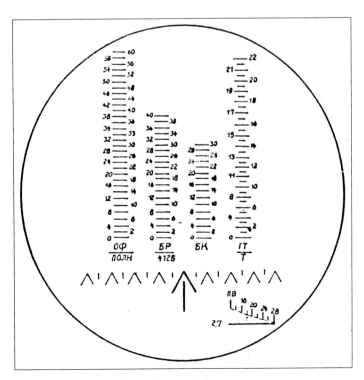

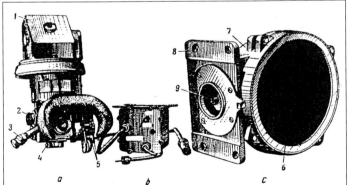

IR-ZF TPN-22-11 with generator and L-2G infrared light: 1. Top; 2. Control button; 3. High-tension attachment; 4. Horizontal adjustment screw; 5. Vertical adjustment screw; 6. Infrared filter; 7. Light housing; 8. Light baseplate; 9. Socket.

Calibration plate for the ZF 2B-22 aiming scope.

Schematics Gallery

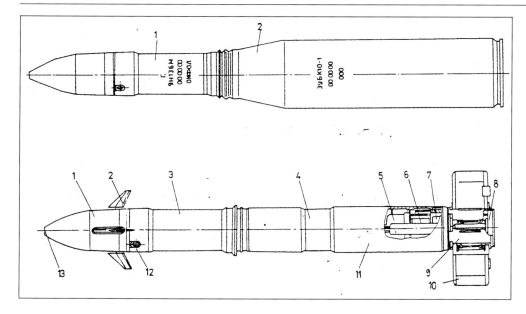

3UBK10-1 100 mm Barrel Rocket (*above*):
1. 9M117 Rocket; 2. Housing.
9M117 Rocket (*bottom*): 1. Rudder drive block; 2. Rudder; 3. Warhead; 4. Marching drive; 5. Generator; 6. Electronic device; 7. Circuit coordinator; 8. Receiver; 9. Stabilizer block; 10. Stabilizer; 11. Apparatus; 12. Switch for light and socket for 9W890 control and testing devices; 13. Air intake.

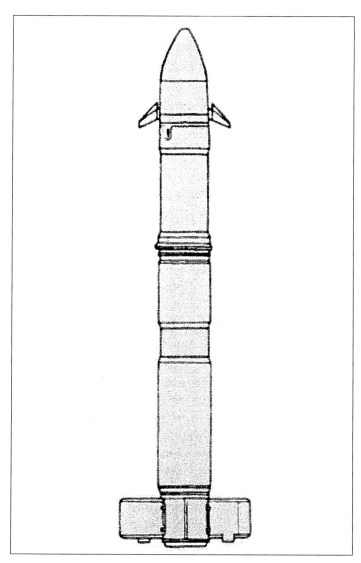

The 9M117 gun-barrel rocket of the UBK-10 100 mm shell of the 9K116 "Bastion" guided missile system.

Shell-fragment Protector with Shield on T-54M, T-55 and T-62 medium tanks: 1. Protector; 2. Cover; 3. Outer glass shield of 1K13 BZ scope block (*see next page*); 4. Lens and photo diode of light protector; 5. Adjusting screw; 6. Frame with glass shield; 7. Shielding cover; 8. Wing screw; 9. Windshield wiper.

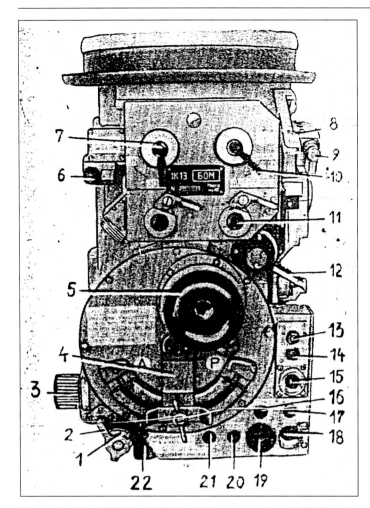

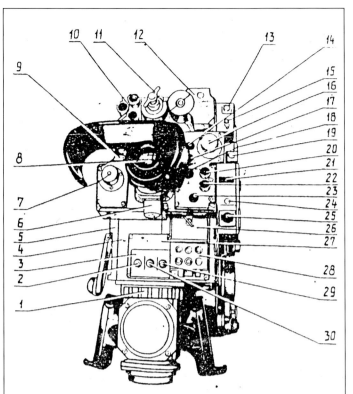

Controls of the 1K13 Optic Block: 1. Distance setting; 2. Calibration adjustment; 3. LIGHT control; 4. Swinging part; 5. Ocular with adjustment; 6. Horizontal adjustment screw; 7. Aiming adjustment sontrol; 8. STAFF lever; 9. Vertical adjustment screw; 10. Loophole; 11. Shield bracket; 12. Curtain; 13. READY control light; 14. OVERHEATING control light; 15. RETURN button; 16. STAFF control light; 17. "U" control light; 18. "U" control switch; 19. Dry cartridge; 20. NIGHT control light; 21. BLOCK HEATING control light; 22. BLOCK HEATING switch.

TDK-K-1 Telescopic Range Finder (front view—T-64B, T-72A and T-80 medium tanks): 1. Adjusting handwheel 2. Tipping button 3. Switchboard 4. Range finder housing 5. Attaching hole for the automatic loading control palen 6. Dioptic setting knob 7. Light setting knob 8. Ocular 9. Front plate 10. Shell-type indicator lights 11. Shell-type switch 12. Closing lid with dry shell 13. Control plug 14. Tipping button for ocular heating 15. Set screw for pump pressure 16. Correction value setting knob 17. Ocular heating indicator light 18. Range-finder tipping button 19. Control plug 20. READY indicator light 21. Numeral indicator 22. SETTING indicator light 23. ATUTO/HAND setting button 24. Close-range target indicator light 25. Close-range target setting button 26. Securing box 27. Indicator panel 28. DRIVE, ENTARR, STABLE, KDT and READY indicator lights 29. DRIVE switch 30. STABLE switch

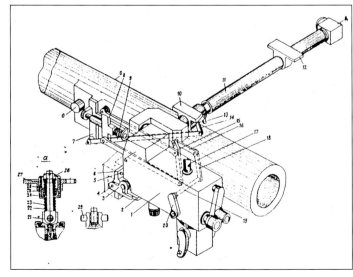

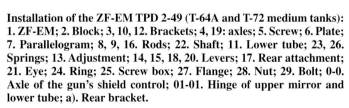

Installation of the ZF-EM TPD 2-49 (T-64A and T-72 medium tanks): 1. ZF-EM; 2. Block; 3, 10, 12. Brackets; 4, 19: axles; 5. Screw; 6. Plate; 7. Parallelogram; 8, 9, 16. Rods; 22. Shaft; 11. Lower tube; 23, 26. Springs; 13. Adjustment; 14, 15, 18, 20. Levers; 17. Rear attachment; 21. Eye; 24. Ring; 25. Screw box; 27. Flange; 28. Nut; 29. Bolt; 0-0. Axle of the gun's shield control; 01-01. Hinge of upper mirror and lower tube; a). Rear bracket.

Schematics Gallery

W-46-6 Engine (T-72, T-72A and T-72B medium tanks), Exhaust Side: 24. Engine attachment; 25. Injection duct; 26. NK-12M Injection pump; 27. Intake manifold; 28. Blower; 29. Upper crankcase; 30. Crankshaft stump; 31. Ducts from the oil pump; 32. Lower crankcase heater; 33. Upper crankcase heater; 34. Cylinder block; 35. Duct for coolant leaving cylinder heads.

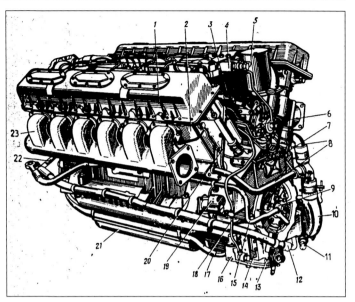

W-46-6 Engine, Intake side: 1. Cylinder head cover; 2. Cylinder head; 3. NK-12M Injection pump rod; 4. Fuel filter; 5. Crankcase vent oil separator; 6. Compressed air distributor; 7. Upper drive cover; 8. Coolant duct to cylinder blocks; 9. Central oil pipe cover; 10. Coolant pump; 11. Coolant outflow plug; 12. Duct from oil pump to MAF oil filter; 13. Engine oil pump; 14. Crankcase vent oil pump; 15. Pipe from MAF oil filter to central oil line cover; 16. Fuel pump; 17. Pipe from oil pump to MZ-1 oil centrifuge; 18. Fuel line from fuel pump to TFK-3 fuel filter; 19. Electric tachometer drive; 20. Crankcase vent duct; 21. Lower crankcase; 22. Pipe from MAF oil filter to blower; 23. Exhaust manifold.

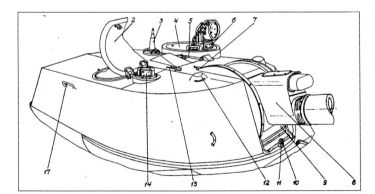

Arrangement of observation devices of the basic 2S1 "Carnation" self-propelled artillery gun; 2, Loader's hatch cover; 3. Antenna base; 4. Commander's cupola; 5. Range finder; 6. Heatable periscope glass; 7. Recuperator filler cap cover; 8. 2A31 Howitzer; 9. Gun mount; 10. Angle limiting screw; 11. Nut; 12. Ventilator cover; 13. PALR carrier; 14. MK-4 periscope; 17. Handhold

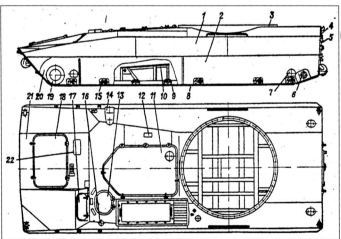

Hull of the basic 2S1 vehicle: 1. Upper sidewall; 2. Lower sidewall; 3. Turret turning crank; 4. Rear wall; 5. Rear door; 6. Rear wheel mount; 7. Attachment for sway damper; 8. Hull bottom; 9. Torsion bar; 10. Engine seat; 11. Engine hatch cover; 12. Air intake ducts; 13. Driver's hatch cover; 14. Exhaust pipe; 15. Torsion bar attachment; 16. Filter vent intake ducts; 17. Windshield cover; 18. Transmission area hatch cover; 19. Reinforcement for power train; 20. Bow of vehicle; 21. Upper bow plate; 22. Ventilator opening.

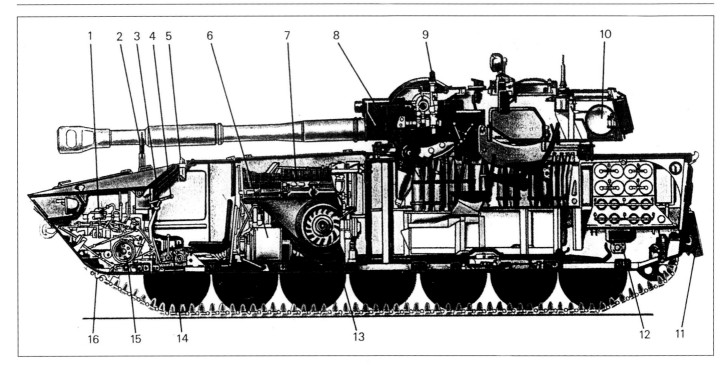

2-S1 "Carnation": 1. Compressed air system; 2. Howitzer attachment for marching; 3. Turning mechanism, clutch and brake levers; 4. Main clutch lever; 5. Optics; 6. Engine warmer; 7. Engine and clutch lubrication system; 8. Ammunition racks; 9. Targeting system; 10. Filter ventilators; 11. Rudder for travel in water; 12. Hydraulic shock absorber; 13. Radiator mantle; 14. Intermediate gears; 15. Main clutch; 16. Starter.

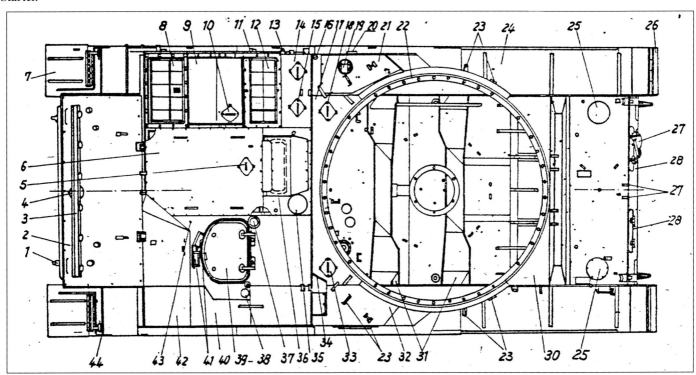

2-S3: 1. Tow hook; 2. Upper bow plate; 3. Dirt deflector; 4. Gearbox cap cover; 5. Fuel cap cover; 6. Engine cover; 7. Front track apron; 8. Air intake grid; 9. Radiator cover; 10. Oil cap cover; 11. Ventilator water drain; 12. Air outlet grid; 13. Air filter water drain; 14. First fuel tank cap cover; 15. Equalizer cap cover; 16. Heating-cooling fuel cap cover; 17. Front cover; 18. Second fuel tank cap cover; 19-21. OW-65G heating-cooling system cover and intakes; 22. Lower turret track attachment; 23. Towrope attachment; 24. Track apron; 25. Rear sway damper cover; 26. Rear track apron; 27. Ammunition hatch; 28. EWZ cover; 29. Sliding shield attachment; 30. Rear cover; 31. Coverings; 32. EWZ compartment; 33. Fourth fuel tank cap cover; 34. Air intake grid for ventilator; 35. Oil filter cover; 36. Air filter cover; 37. Engine-room air intake; 38. Driver's compartment ventilator intake; 39. Driver's hatch cover; 40. Driver's compartment cover; 41. Periscope shafts; 42. Slanted covering plate; 43. Windshield washer cover; 44. Front track apron torsion bar.

Schematics Gallery

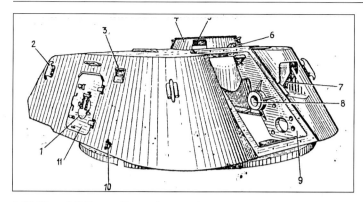

2-S3 "Acacia" Turret (front view): 1. Shell ejection hatch; 2. Hook; 3. Gunner's scope shaft; 4. Commander's cupola; 5. Commander's scope shafts; 6. Staff bracket; 7. OP5-38 scope opening; 8. Shield attachment; 9. Gun mount; 10. Gunner's hatch latch; 11. Gunner's hatch.

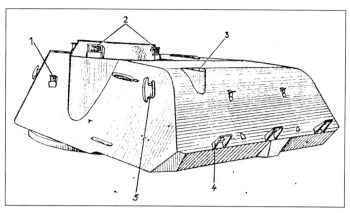

2-S3 Turret (rear view): 1. Gunner's scope shaft; 2. Commander's scope shafts; 3. Antenna bracket; 4. Canvas cover attachments; 5. Hook.

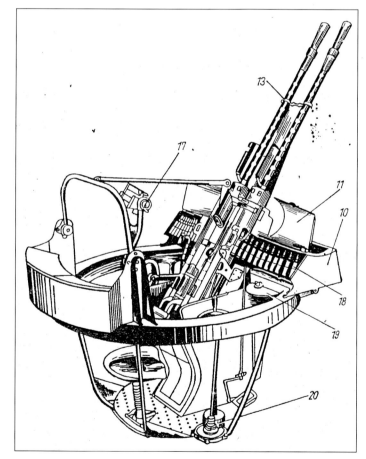

ZPTU-2 self-propelled AA M.g. (on BTR-40 and BTR-152) Overall view of gun mount (from right rear): 17. WK-4 anti-aircraft aiming control; 18. Ammunition belt; 19. Side braking device; 20. Traversing gear.

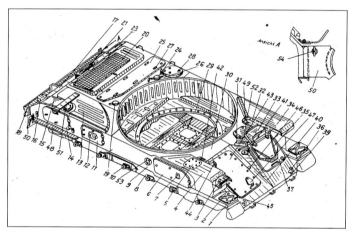

ZSU-23-4W1 and ZSU-23-4M self-propelled anti-aircraft gun (as of 1971 model): 1. Front track apron; 2. Tool-chest lid; 3. Headlight protector cap; 4. Generator air intake; 5. Transformer access hatch; 6. Transformer hatch; 7. Electrically operated transformer air vent; 8. Lower side plate; 9. Upper side plate; 10. Swinging arm mounts; 11. Generator hatch; 12. Generator air vent; 13. Gas-turbine air-filter air intake, opens when starting engine; 14. Gas turbine hatch; 15, 26. Powerplant armor plates; 16. Gas turbine exhaust; opens automatically when starting engine; 17. Upper rear plate; 18. Ejector cover; 19. Rear fuel filler hatch cover; 20. Air intake with louvers; 21. Tool chest; 23. Fire extinguisher hatch cover; 24. Oil filler hatch cover; 25. Engine cover; 27. Lock; 28. Air filter hatch cover; 29. Ring for attaching turret turning circle; 30. Front hull covering plate; 31. Driver's compartment air vent; 32. Spring equalizer; 33. Observation device cover; 34. Engine air exit shaft; 35. Windshield cover; 36. Front tow hook; 37. Driver's visor cover; 39. Lower bow plate; 40. Right side plate of driver's compartment; 41. Obersvation device; 42. Bettery gas exit cap; 43. Fuel tank hatch cover; 44. Fuel filler hatch cover; 45. Upper bow plate; 46. Chassis number plate; 47. Windshield washer fluid hatch cover; 48. Folding shield on sidewall; 49. Spring equalizer shaft; 50. Tow-bar bracket; 51. Gas turbine access hatch; 52. Driver's emergency exit hatch; 53. Electric connection cover (SchRWP); Ground wire wingnut.

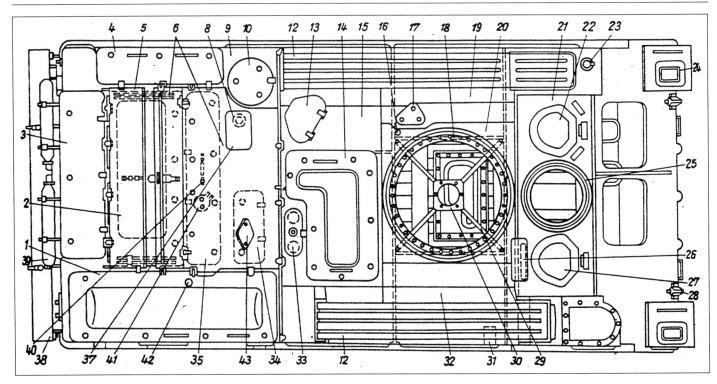

1. Generator area cover; 2. Ejector cover; 3. Rear plate; 4. Heater hatch cover; 5. Folding cover plate; 6. Folding cover plates; 8. Oil filler cap screw; 9. Rear cover plate; 10. Engine air filter cover; 12. Air shafts; 13. Air exit hatch cover; 14. Generator hatch cover; 15. Central cover plate; 16. 1A18M1 system dust ejector hatch cover; 17. Fighting compartment air intake shaft; 18. Circular turning rail; 19, 32. Left and right cover plates; 20. Cover plate; 21. Front cover plate; 22, 27. Cover over driver's and mechanic's seats; 23. Windshield washer fluid filler cover; 24. EWZ box; 25. Commander's cupola turning rail; 26. Fighting compartment ventilator cover; 28. Tow hook; 29. Split-ring carrier baseplate console; 31. Battery area ventilator cover; 33. Left and right front fuel tank filler cover; 34. Rear fuel tank filler cover; 35. Engine cover; 37. Oil filler cover in folding panel; 38. Tow bar; 39. Compressed air tank; 40. Attachment for hand light; 41. Top light; 42. Gas turbine oil dipstick cover; 43. Armor plate over rear fuel filler and dipstick.

The 8U218 launching vehicle with 3R-1 "Owl" tactical rocket; NATO code Frog-1.

Schematics Gallery

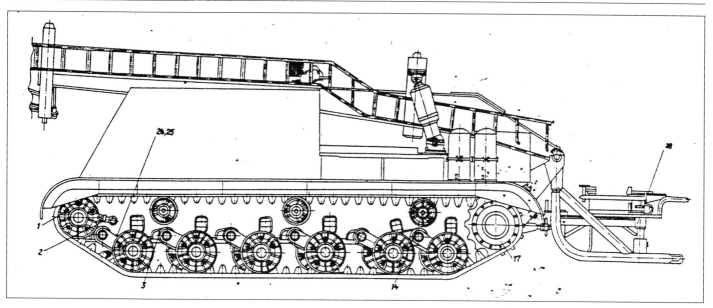

The 8U218 launching vehicle.

The 2P19 launching vehicle with 8K11 operative-tactical rocket complex, NATO code SS-1/SCUD-A.

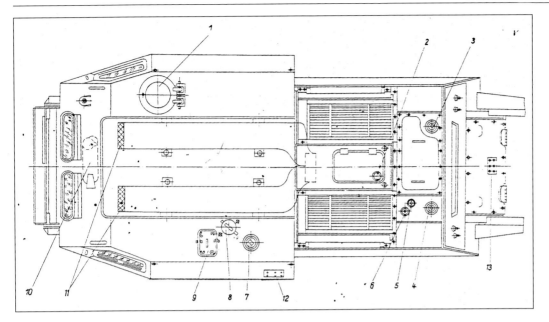

Covers and Set Screws in the Hull (seen from above) on the 8U218 and 2P19 launchers: 1. Panoramic viewing hatch; 2. Set screw for main clutch lubrication; 3, 4. Fuel filler set screws; 5. Oil tank dipstick; 6. Oil filler set screw; 7. Hydraulic system oil tank set screw; 8. Electric system ventilator opening; 9. Coolant filler opening; 10. Ventilator; 11. Air intake shafts; 12. Opening to check generator brushes in electric cystem; 13. Planetary gear lubrication opening (positions 8, 10, 11 and 12 apply only to the 2P19 device).

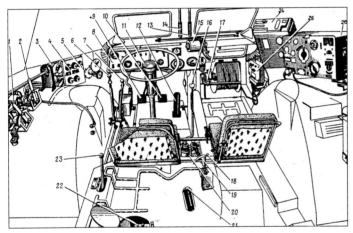

Controls in the BRDM-2: 1. Hydraulic system vent for additional wheels; 2. Hydraulic system vent for caps and deflectors; 3. Ventilation flap crank; 4. Air divider block; 5. Automatic main vent for tire-pressure regulator; 6. Shift lever for road and off-road gears; 7. Shift lever for front-wheel drive; 8. Instrument panel; 9. Clutch pedal; 10. Brake pedal; 11. Horn button; 12. Gas pedal; 13. Lever operating commander's hatch panel; 14. Lever operating driver's hatch panel; 15. Windshield wiper motor; 16. Gearshift lever; 17. Winch clutch pedal; 18. Hand throttle; 19. Hand brake lever; 20. Air flap button; 21. Water propulsion shift lever; 22. Winch shift lever; 23. Shift lever for additional wheels; 24. Navigation coordinate system; 25. Dosing gauge; 26. Radio.

A BRDM-2 of the 4th Infantry Regiment of the U.S. Army at Hohenfels in 1988.

Right: The BRDM-2U command vehicle after the disbanding of the NVA at Grossenhain, Saxony.

Schematics Gallery

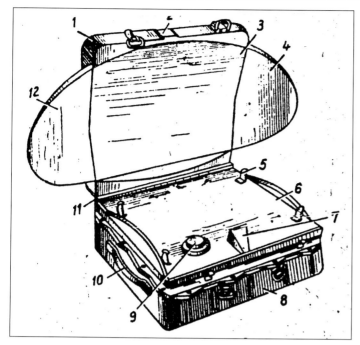

Sender-Receiver on the BRM-1K armored reconnaissance vehicle: 1. Upper lid; 2. Control; 3. Reflector; 4, 12. Reflector segments; 5. Brackets; 6. Baseplate; 7. Amplifier; 8. Lower lid; 9. Lid for frequency changing switch; 10. Carrying handle; 11. Spring. *Right:* Another view of the Sender-Receiver: 1. Lid; 2. Reflector segment; 3. Cable; 4. Housing; 5. Sender-receiver; 6. Flange; 7. Screw; 8. Wire.

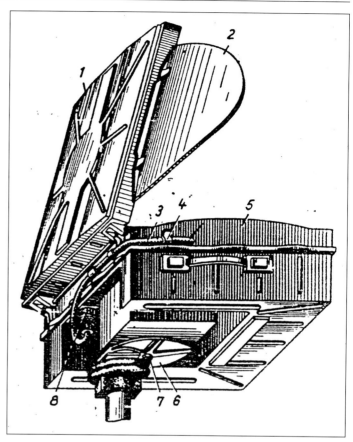

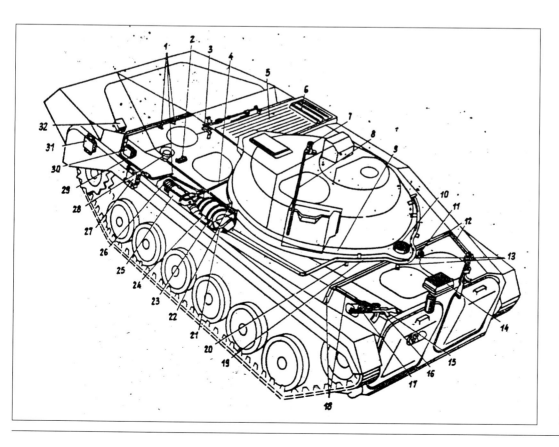

BRM-1K KCB Protection System: 1, 13, 20. Fresh air duct ends; 2. Engine removal system; 3. Blind and ejector flap controls; 4. Dust removal system; 5. Ejector blind; 6. Ejector flaps; 7. Turret ventilator; 8. Cartridge gas ejector; 9. Ring air duct; 10. Engine air duct for water travel; 11. Crew area ventilator; 12, 18. Main ventilator duct; 14. Crew area heater; 15 Machine pistol; 16. Gas collector; 17. Machine pistol ball mantlet; 19. Main air duct; 21. Heater; 22. Air duct from ring air ducting; 23. FPT filter; 24. Filter bracket; 25. Blower; 26. Fan dust remover; 27. Nuclear and chemical sensing system; 28. Cyclonic filter; 29. Temperature heater for nuclear and chemical sensing system regulator; 30. Nuclear and chemical sensor indicator; 31. Nuclear and chemical sensor power source; 32. KR-40 relay box.

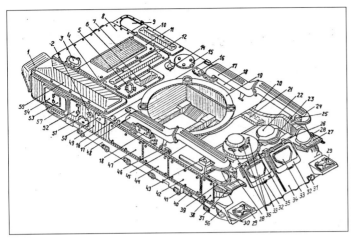

1S91M1 and 1S91M2 reconnaissance and guiding station: 1. Gas turbine air intake shield; 2. Outside power plug cover; 3. Rear fuel filler cover; 4. Removable outside plate; 5. Front fuel filler cover; 6. Front fuel tank cover; 7. Grid; 8. Rear fuel tank cover; 9. Tool chest; 10. Heater cover; 11. Air intake; 12. Engine cover; 14. Oil filler cover; 15. Air filter cover; 16. Technical plate; 17. Air intake shaft; 18. Air exit; 19. Body support ring; 20. Fighting compartment cover; 21. Ventilator opening; 22. Observation devices cupola; 23. Observation devices shield; 24. Fighting compartment air intake shaft; 25. Windshield washer fluid filler cap; 26. Observation port shaft; 27. Headlight covers; 28. Radio antenna flange; 29. EWZ box; 30. Track shield; 31. Front tow hook; 32, 33. Hatch covers; 34. Bow plate; 35. Commander's cupola; 36. Observation devices shield; 38, 42, 44, 47. Tool cabinet lids; 39. Right front assembly hatch; 40. Hydraulic shock absorber shaft; 41. Swinging arm box; 43. Air passage from blower to powerplant; 45. Planking attachment; 46. Hand control attachment; 48. Lower right sidewall; 49. Upper right sidewall; 50. SVA hatch cover; 51. SVA fuel valve cover; 52. Generator ventilation air exit; 53. Gas turbine exhaust gas exit; 54. Gas turbine compartment hatch; 55. Gas turbine compartment cover; 56. Leading wheel bay; 57. Drive train bay.

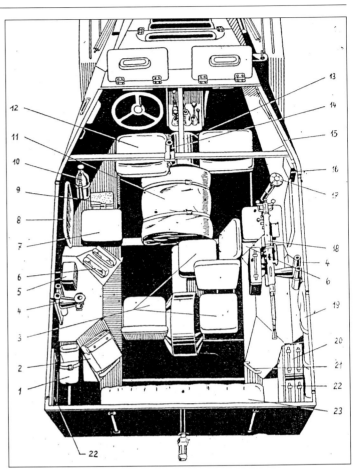

BTR-40 armored transporter, Driver's and Fighting Compartments: 1. Machine-gun tool kit; 2. Hand grenade compartment; 3. Folding crew seat; 4. Side attachment for machine gun; 5. Blocks of synthetic gas; 6. Hand grenade rack; 7. Crew seat; 8. Backrest of crew seat; 9. Tool kit; 10. OU-2 fire extinguisher; 11. Canvas cover storage; 12. Driver's seat; 13. Spare oil tank; 14. Movable machine gun attachment; 15. Antenna bracket; 16. First aid kit; 17. Rear fuel filler cap; 18. Machine gun in marching position; 19. Spare machine gun barrel; 20. Ammunition belt box; 21. Handhold; 22. Windshield bracket; 23. Bench for three crewmen.

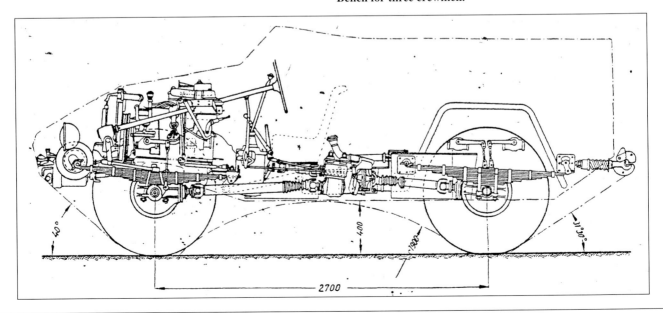

Schematics Gallery

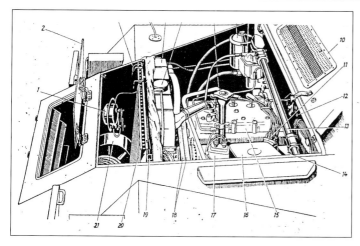

BTR-40 1. Horn; 2. Forward hatch cover; 3. Radiator louver opener; 4. Radiator; 5. Radiator cap; 6. Oil filler cap; 7. Oil filter; 8. Accelerator; 9. Electric wires; 10. Radio grid; 11. Distributor; 12. Coil; 13. Air filter duct; 14. Anti-interference device; 15. Engine; 16. Regulator switch; 17. Generator; 18. Ventilator-generator belt; 19. Oil cooler; 20. Radiator louvers; 21. Winch.

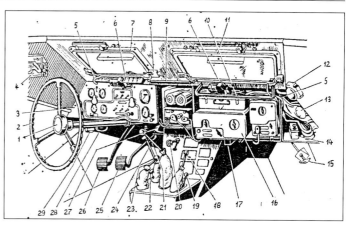

BTR-40 Operating Levers, Control and Indicator Devices: 1. Horn button; 2. Steering wheel; 3. Instrument panel; 4. Visor shield; 5. Water drainpipe; 6. Windshield wiper; 7. Water drain; 8. Windshield plate bolt; 9. Water drainpipe; 10. Handhold; 11. Head protector; 12. Radioman's light; 13. Head protector storage; 14. Morse key; 15. Radiator louver handle; 16. Radio; 17. Radio tool kit; 18. All-wheel drive shift lever; 19. Winch lever; 20. Starter button; 21. Gearshift; 22. Foot pedal; 23. Road-off-road shift lever; 24. Hand brake; 25. Foot brake; 26. Clutch pedal; 27. Oil can; 28. Air filter; 29. Electric anti-interference shield.

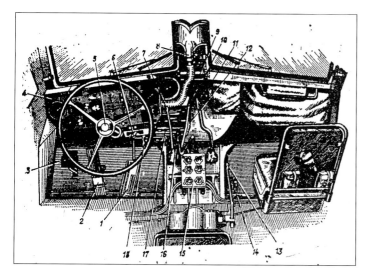

BTR-152W1 Controls (driver's compartment): 1. Gas pedal; 2. Brake pedal; 3. Clutch pedal; 4. Instrument panel; 5. Horn button; 6. Tire-pressure regulator vent lever; 7. Divider gearshift; 8. Hot air exit; 9. Windshield wiper switch; 10. Gearshift; 11. Hand brake; 12. Heater; 13. Subsidiary drive shift lever; 14. Battery box; 15. Tire pressure regulator vent; 16. Gearshift; 17. Heater cover; 18. Radiator louver control handle.

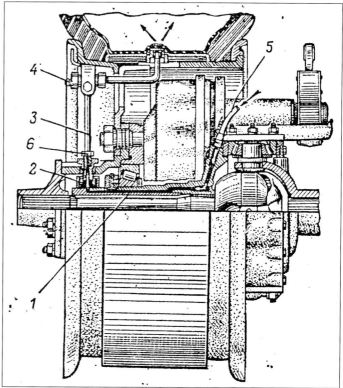

BTR-152W1 and BTR-152U Air ducting through the axle shank pin: 1. Air passage canal; 2. Air intake head; 3. Air duct; 4. Blocking valve; 5. Air hose; 6. Brace.

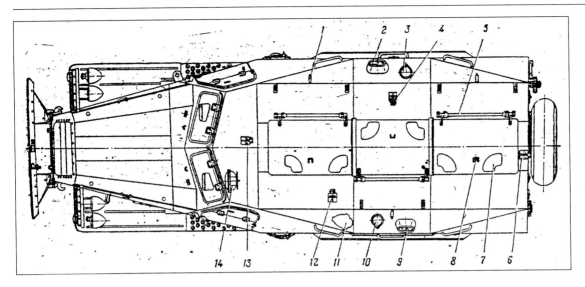

BTR-152K: 1. Bar to hold hatch cover open; 2, 9. Observation devices; 3, 10. Close-combat loopholes; 4, 6, 12, 13. Machine-gun brackets; 5. Hatch-cover hinge; 7. Hatch cover; 8. Hatch-cover rest; 11. Hull ventilator protector; 14. Orifice for TWN-2 night-vision device.

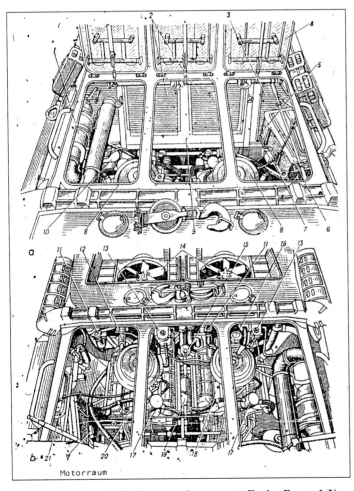

BTR-60PA, PB and BTR-70 armored transports, Engine Room: 1. Ventilation filter; 2. Engine-room bulkhead cover with latches; 3. Starter battery cover; 4. Engine-room cover; 5. Starter batteries; 6. Battery cover; 7. Electric tension meter; 8. Air filter; 9. Removable part of engine-room bulkhead; 10. Hydraulic system container; 11. Engine coolant fillers; 12. Fan-belt tension spring.

Controls of the BTR-60PB: 1. Air-flap and throttle-flap buttons; 2. Radiator cover lever; 3. Power source for driver's night-vision device; 4. Water drive vent closing cap lever; 5. Front light plug; 6. Air divider block; 7. Horn button; 8. Steering wheel; 9. Lever for driver's windshield cover; 10. Windshield wiper switch; 11. Windshield wiper lever; 12. Gear selector switch; 13. Wave deflector lever; 14. Gas pedal; 15. Hand brake lever; 16. Shift lever for 1st and 2nd axles; 17. Road/off-road shift lever; 18. Winch lever; 19. Water drive switch; 20. Compressed air drain hatch; 21. Brake pedal; 22. Clutch pedal; 23. Driver's compartment heater.

Schematics Gallery

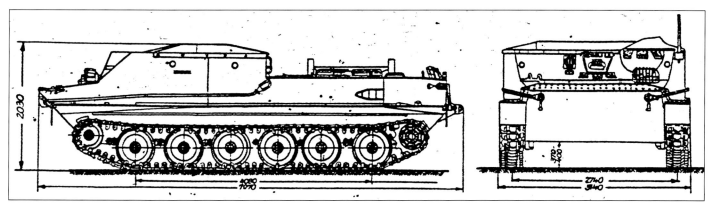

BTR-50P armored transport: 1. Lower bow plate; 2, 4 & 20. Tow hooks; 3. Wave deflector; 5. Middle bow plate; 6. Upper bow plate; 7. Driver's hatch cover; 8. Back wall; 9. Right front covering plate; 10. Folding rear wall; 11. Fuel-tank covering plate; 12, 19 & 29. Handholds; 13. Engine cover; 14. Folding rear wall; 15. Power-train covering plate; 16. Radiator cover; 17. Ejector opening; 18. Rear covering plate; 21. Rear side plate; 22. Water drain opening; 23. Swinging-ari mount block; 24. Lower side plate; 25. Middle side plate; 26. Close-combat loophole; 27. Compensating gear; 28. Swinging-arm limiting brackets; 30. Hydraulic shock absorber attachment; 31. Commander's cupola; 32. Mount for eccentric; 33. Front side plate; 34. Wave-breaker activator box.

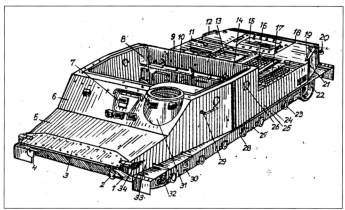

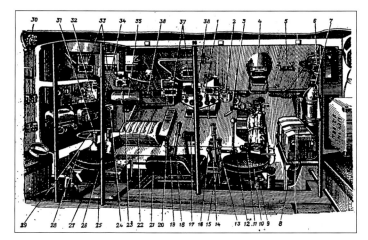

BTR-50PK armored transport: 1. Driver's instrument panel; 2. GPK-59 course indicator; 3. Wave-breaker handwheel; 4. TPKU-28 observation device for group leader of motorized riflemen; 5. Observation device bracket; 7. Loophole; 8. Starter battery; 9. Battery switch; 10. Group leader's seat; 11. Fuel pump; 12. Gearshift lever; 13 & 18. Steering levers; 14. Gas pedal; 15 & 20. Water propulsion steering levers; 16 & 19. Divider drive shift levers; 17. Clutch pedal; 21. Driver's seat; 22. Machine-gun bullet belt; 23. Regulator switch; 24. Electric filter chain; 25. Tool box; 26. Commander's seat; 27 & 31. Fire extinguishers; 28. Signal pistol box; 29. Engine hour counter; 30. Filter ventilator; 32. Radio; 33. Commander's observation device; 34. AS-2 automatic switch; 35. Compressed air tank; 36. Securing box; 37. Driver's observation device; 38. Driver's hatch cover.

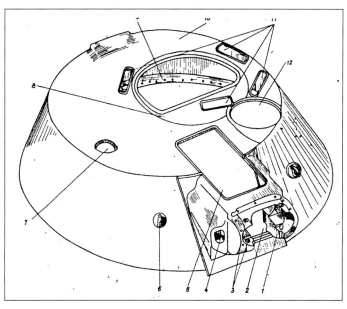

BMP-1 infantry combat vehicle turret: 1. Gun mount opening; 2. Tank gun opening; 3. Openings for breeches; 4. Machine-gun opening; 5. PALR loading hatch; 6. Hook; 7. Ventilator opening; 8. Aimer's hatch; 9. Covering; 10. Turret top; 11. Openings for observation devices; 12. Opening for targeting scope.

BMP-1 Infantry Combat Vehicle

1. Hull
2. Wave deflector
3. Deckplate
4. Instrument panel
5. Driver's optics
6. Driver's hatch release
7. Engine
8. Louver lever
9. Air filter
10. Commander's seat
11. Tank gun
12. PALR starter
13. PALR
14. Turret
15. Headlight
16. Right gunner's seat
17. Fuel tank
18. Riflemen's seats
19. Battery
20. One-man AA rocket
21. Rear hatch
22. Rear combat hatch
23. Tracks
24. Road wheel
25. Torsion bar
26. Engine bulkhead
27. Driver's seat
28. Clutch pedal
29. Gas pedal
30. Generator
31. Leading wheel
32. Return roller
33. Loopholes
34. Observation device
35. Riflemen's optics
36. Steering lever
37. Gearshift lever
38. Track tension device

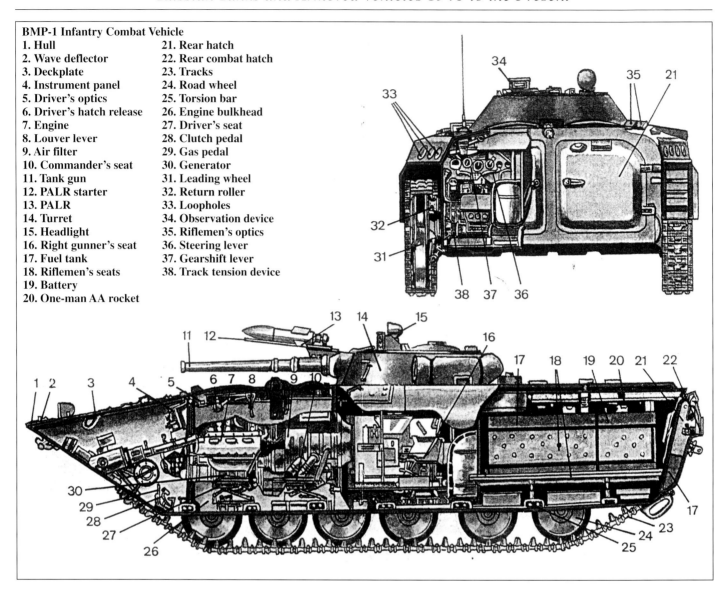

BMP-1 and BMP-1P infantry combat vehicles: 1. Swinging-arm mount; 2. Lower side piece; 3. Driveshaft flange; 4. Upper side piece; 5. Upper bow plate; 6. Hoop; 7. Ribbed bow plate; 8. Driver's hatch cover; 9. Closing screw over gearbox oil filler; 10. Commander's hatch cover; 11. Engine hatch cover; 12. Closing screw over cooling system equalizer; 13. Closing screw over lubrication filler; 14. Louvers; 15. Ejector flap; 16. Bulkhead; 17. Upper hull; 18. Bottom of hull; 19. Air intake pipe; 20. Crew's hatch cover; 21. Observation device shaft; 22. Rear part; 23. Rear door with fuel tank; 24. Closing screw for fuel filler; 25. Antenna base; 26. Rear tow hook; 27. Flange for leading-wheel eccentric; 28. Machine-pistol loophole covers; 29. Swinging-arm limiter; 30. Bending limiter of first swinging arm; 31. Hydraulic swing damper's upper bracket; 32. Machine-gun hatch cover plate; 33. Return roller flange; 34. Cover of FPT-200M filter; 35. Handhold; 36. Deflector; 37. Rear door lock button; 38. Buffer; 39. Filter air intake cover; 40. Air exit vent cover; 41. Air ejector vent cover.

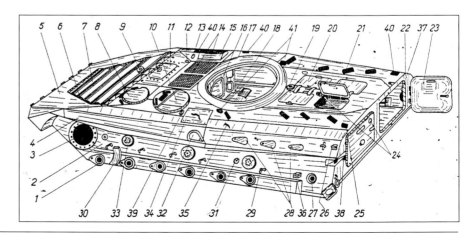

Schematics Gallery

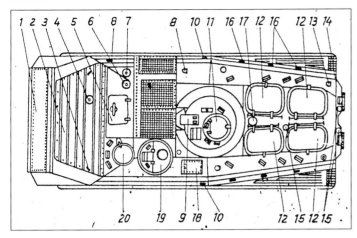

Openings and Hatches in the upper part of the BMP-1 and BMP-1P bodies: 1. Wave deflector; 2. Ribbed bow plate; 3. Handle; 4. Closing screw for gearbox oil dipstick opening; 5. Hatch cover over engine; 6. Closing screw for coolant filling opening; 7. Closing screw for oil tank filler opening; 8, 13. Bilge pump drainpipe opening; 9. PALR loading hatch (only BMP-1); 10. Machine-gun loophole; 11. Turret hatch; 12. Crew space hatch cover; 14. Rear door; 15. Fuel tank filler cover; 16. Loophole; 17. Air intake duct; 18. Access to absorption filter; 19. Commander's hatch cover; 20. Driver's hatch cover.

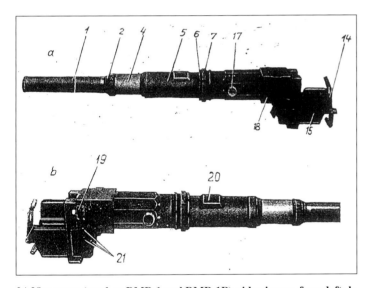

2A28 cannon (used on BMP-1 and BMP-1P), side view: a. from left; b. from right: 1. Barrel; 2. Front nut; 4. Spacer mantle; 5. Outer cylinder of recoil brake; 6. Rear nut; 7. Mount attaching ring; 14. Shell ejector; 15. Ejector; 17. Shield attachment with bearing; 18. Cradle; 19. TKB screw connector; 20. Plate with boreholes; 21. MG rack attachment screwholes.

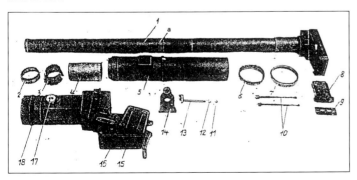

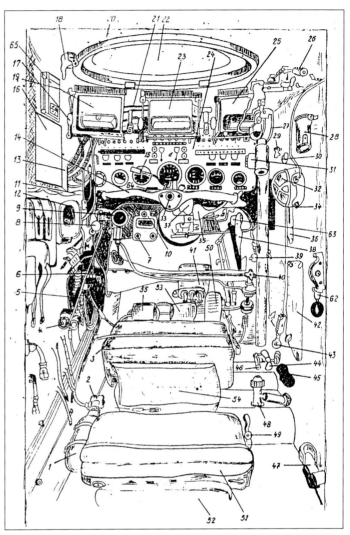

BMP-1 and BMP-1P Driver's Compartment: 1. Compressed air flask; 2. Compressed-air flask valve; 3. Driver's seat; 4. Wave-deflector, air intake and dust ejector handle; 5. Electropneumatic valve; 6. Rod; 7. Course indicator control button; 8. Wave-deflector control; 9. Compressed-air gauge; 10. Course indicator; 11. High-tension cable for drivert's night-vision device; 12. A-1 speaker system; 13. R 123 radio power generator; 14. Tiller; 15. Road/off-road control lever; 16. R 123 radio set; 17. Periscope attachment; 18. Driver's hatch cover latch; 19. Left periscope for driver; 20. Driver's compartment ceiling; 21. Air duct from ventilation filter; 22. Driver's hatch cover; 23. Central periscope for driver; 24. Driver's instrument panel; 25. Right periscope for driver; 26. Louver and ejector flap control handle; 27. Driver's hatch-cover opening handle; 28. Upper part of machine-pistol bracket; 29. Engine-room access hatch cover; 30, 36, 42. Engine-room access hatches; 31. Instrument lights; 32. Light; 33. Horn button; 34. Light switch; 35. Gearshift lever; 37. Steering column; 38. Brake handle; 39. Heater control; 40. RPM control; 41. KR-40 relay box; 43. Engine-cooling control; 44. Heater fuel valve; 45. Fuel valve; 46. Oil-water separator control; 47. Lower machine-pictol bracket; 48. Coolant release valve; 49. Seat adjustor handle; 50. Gas pedal; 51. Commander's seat; 52, 54: Headgear compartments; 53. Brake pedal; 55. Clutch pedal; 62. Heater air-flap lever; 63. Hand-brake lever; 64. Red "release hand brake" light; 65. Side periscope for driver (in SPWs built as of July 1975).

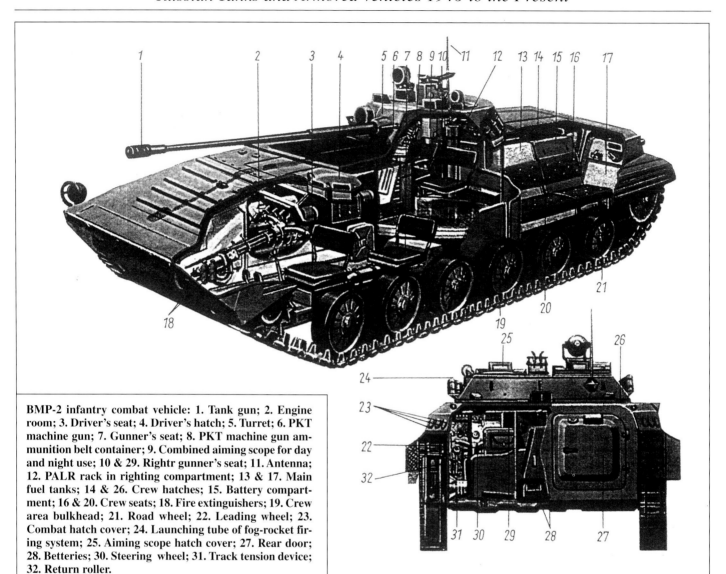

BMP-2 infantry combat vehicle: 1. Tank gun; 2. Engine room; 3. Driver's seat; 4. Driver's hatch; 5. Turret; 6. PKT machine gun; 7. Gunner's seat; 8. PKT machine gun ammunition belt container; 9. Combined aiming scope for day and night use; 10 & 29. Rightr gunner's seat; 11. Antenna; 12. PALR rack in righting compartment; 13 & 17. Main fuel tanks; 14 & 26. Crew hatches; 15. Battery compartment; 16 & 20. Crew seats; 18. Fire extinguishers; 19. Crew area bulkhead; 21. Road wheel; 22. Leading wheel; 23. Combat hatch cover; 24. Launching tube of fog-rocket firing system; 25. Aiming scope hatch cover; 27. Rear door; 28. Betteries; 30. Steering wheel; 31. Track tension device; 32. Return roller.

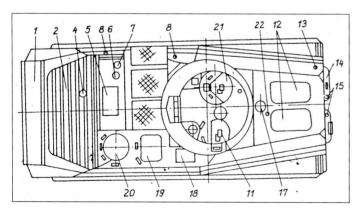

BMP-2: 1. Wave deflector; 2. Ribbed bow plate; 3. Handle; 4. Closing screw for gearbox oil dipstick opening; 5. Engine hatch cover; 6. Closing screw for coolant filling opening; 7. Closing screw for oil tank filler opening; 8, 13. Bilge pump drainpipe opening; 11. Right gunner's hatch cover; 12. Crew area hatch cover; 14. Rear door; 15. Fuel filler cover; 16. Loophole; 17. Air intake duct; 18. Absorption filter access; 19. Hatch cover; 20. Driver's hatch cover; 21. Commander's hatch cover; 22. Fuel tank filler cover.

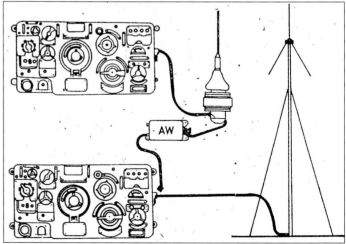

Attachment of the 4-meter staff antenna and the 10-meter semi-telescopic mast on command vehicles.

Schematics Gallery

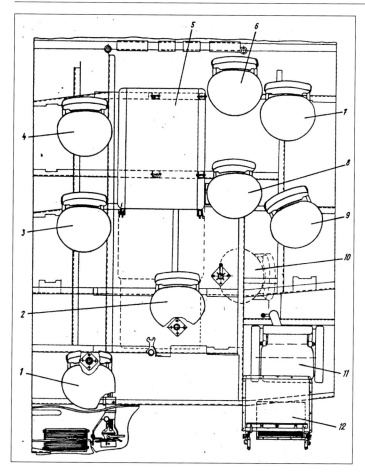

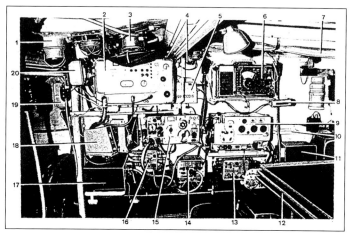

BTR-50PU, Right sidewall of the workspace: 1, Antenna attachment for rod antenna of R 111 UKW radio; 2. R 111 sender-receiver; 3. A2 component of R 124 speaker system; 4. Antenna attachment for rod antenna of R 130 or R 123M radio; WSU-T tuner; 6. R 311 receiver; 7. 4-m rod antenna; 8. WSU-T remote control; 9. R 130 sender-receiver; 10. Attachment for antenna equivalent of R 130 KW radio; 11. BP-UM-26 (R 130) radio generator; 12. Worktable; 13. R 130/R 311 control panel; 14. BP-26 (R 123M) radio generator; 15. R 123M sender-receiver; 16. R 111/R 123 control panel; 17. R 111 UKW radio generator; 18. Antenna attachment for R 111 UKW radio; 19. A1 component of R 124 speaker system; 20. UPU-105U amplifier for R 111 UKW radio.

Below: BTR-50PU comand-staff 1. Wave breaker; 2, 35. Track aprons; 3, 34. Headlight protectors; 4. Driver's hatch cover; 5, 14, 31. Handholds; 6. Antenna attachment for R-403 BM radio set; 7. Armored cap of ventilator (air intake); 8. Front cover plate of workspace; 9. Turning cupola; 10. Antenna attachment for R-112 or R-113 radio; 11. Right entrance hatch; 12. Rear cover plate of workspace; 13. Aromred cap of air centrifuge (thickener); 15, 29. Steps; 16. Ladder; 17. Engine cover; 18. Right cover plate; 19. Armored fuel-tank box; 20. Gearbox cover plate; 21. Radiator grille; 22. Air exit grille; 23. Telescopic mast; 24. Armored cap of air exit (ejector); 25. Air ducts to engine; 26. Left entrance hatch; 27. Anetnna attachment for R-105 radio; 28. Upper light hatch; 30. Antenna attachment for R-105 U radio; 32. Commander's hatch; 33. Upper bow plate.

BTR-50PU command-staff vehicle, arrangement of workplaces in the work space: 1. Commander's seat; 2. First officer's seat; 3, 4 & 7. Radiomen's seats; 5. Worktable; 6. Third officer's seat; 8. Second officer's seat; 9. Radio troop leader's seat; 10. First officer's seat with worktable pulled out; 11. Commander's seat; 12. Commander's table.

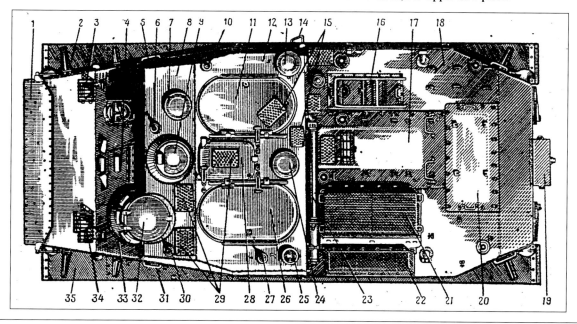

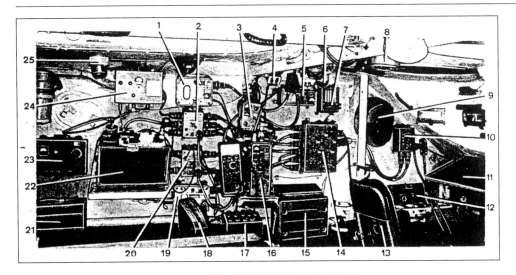

BTR-50PU, Left sidewall of the workspace: 1. UPU-105U amplifier; 1. R 105M comtrol panel; 3. UM amplifier; 4. Generator switch (generator for UM amplifier and radio power); 5. Antenna attachment for R 105M UKW radio; 6. Antenna attachment for rod antenna of R 105M UKW radio; 7. Loudspeaker; 8. A2 component of R 124 speaker system; 9. Pocket for commander's headgear; 10. Commander's control panel; 11. Commander's worktable; 12. Field phone; 13. Commander's seat; 14. Radio controls; 15. P 193A telephone controls; 16. R 105M UKW radio; 17. Generator for P 193A telephone; 18. First officer's seat; 19. UPU-107 amplifier; 20. R 107 control panel; 21. Box for spare batteries (KN 14 - R 105M, 2KNP-20 - R 107); 22. R 107 UKW radio; 23. Electric controls; 24. Antenna attachment for R 107 radio and R 311 receiver; 25. Antenna attachment for rod antenna of UKW radio or R 311 receiver.

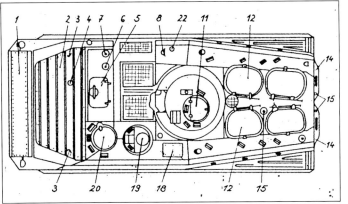

Left: BMP-1KSch command-staff vehicle: 1. Wave deflector; 2. Ribbed bow plate; 3. Handhold; 4. Closing screw of opening for gearbox oil dipstick; 5. Engine jatch cover; 6. Closing screw of opening for coolant filler; 7. Closing screw of opening for oil tank filler; 8, 13. Bilge-pump drain opening; 10. Machine-gun loophole; 12. Crew area hatch cover; 14. Rear hatch; 15. Fuel-tank filler hatch cover; 16. Loophole; 17. Air exit duct; 18. Absorption filter access hatch; 19. Hatch cover; 20. Driver's hatch cover; 22. Heating and ventilation system exhaust opening.

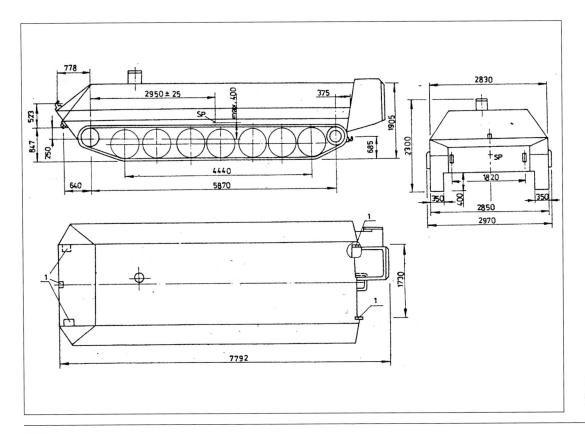

KSchM-9S743 command-staff vehicle with R-330P VHF inerference set.

Schematics Gallery

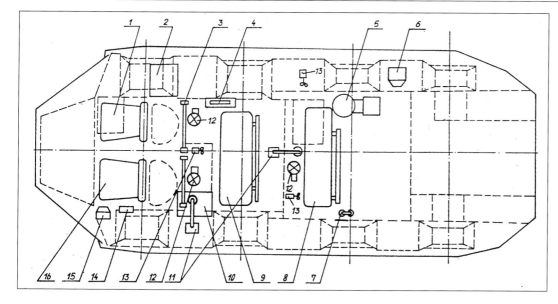

Extra equipment of R145BM command-staff vehicle: 1. Commander's seat; 2. Wall closet for documents; 3. Sunshade; 4. PO-1 worktable; 5. Filter ventilator; 6. Heater for commander's and equipment space; 7. Fire extinguisher; 8. Radioman's seat; 9. Seat; 10. Worktable; 11. Workplace lights; 12. Ceiling lights; 13. Ventilators; 14. First-aid kit; 15. Heater for driver's area; 16. Driver's seat.

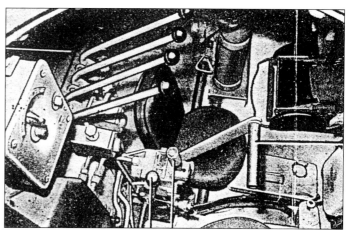

Above and Left: BTS-3 medium armored towing tractor, seat at left next to commander's seat: 3. Winch clutch-brake control lever; 4. Winch power control lever; 5. Lever to set RPM by commander's seat; 6. Crane control panel.

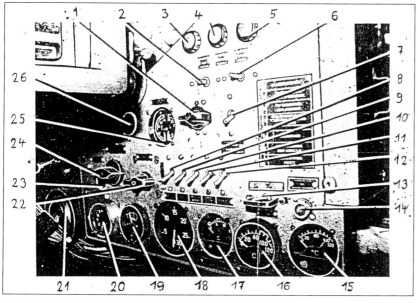

BTS-3 driver's instrument panel: 1. Outside light switch; 2. Left blinker switch; 3. Left blinker indicator light; 4. Indicator light for border lights; 5. Right blinker indicator light; 6. Right blinker light switch; 7. All-around light switch; 8. Clock heater switch; 9. Course indicator switch; 10. Bilge-pump switch; 11. Infrared headlight switch; 12. Power switch for night-vision device (driver); 13. Foglaying switch; 14. Driver's periscope washer switch; 15. Coolant thermometer; 16. Oil thermometer; 17. Oil manometer; 18. RPM indicator; 19. Electric tension gauge; 20. Clock; 21. Horn for "SIGSTA" signal system; 22. Electric oil pump switch; 23. Starter switch; 24. RPM indicator switch; 25. Fuel gauge; 26. Horn button.

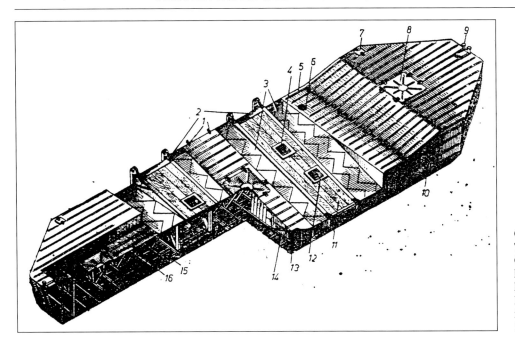

GSP-55 self-propelled ferry: 1. Ducts; 2. Track supports; 3. Metal panels; 4, 12. Openings; 5. Deck; 6. Bilge-pump drain hose attachment opening; 7. Drainage hatch; 8. Entry hatches; 9. Bollard; 10. Drain cover; 11. Planking; 13. Corner iron; 14. Material hard to sink; 15. Floor; 16. Transverse frame.

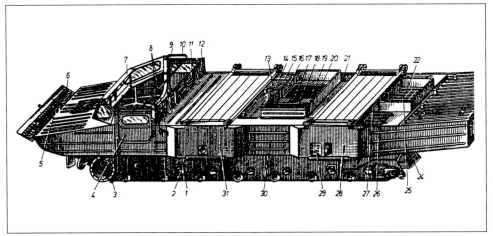

GSP-55: 1. Swinging-arm limiter; 2. Plates; 3. Track deflector; 4. Door; 5, 23. Bollards; 6. Wave deflector; 7. Windshield frame; 8. Bilge-pump pipe support; 9. Door bolt; 10. Bow; 11. Bulkhead; 12. Hatch cover; 13. Plate; 14. Cover plate; 15, 21, 26, 27, 29. Covers; 16. Apron; 17. Cutout; 18. Folding wall panel; 19. Spring; 20, 22. Grilles; 24. Dust flap; 25. Ventilator shaft; 28. Rear frame member; 30. Pressed steel profile; 31. Front frame member.

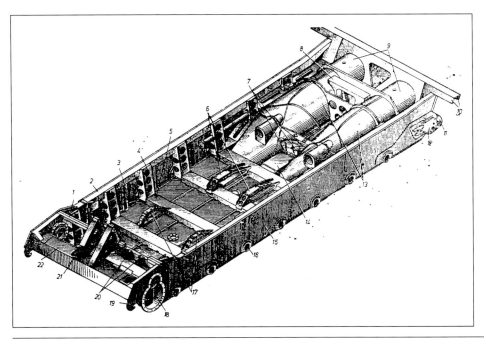

GSP-55: 1, 5, 22. Transverse frame members; 2, 14. Longitudinal frame members; 3, 4. Supports; 6, 7. Engine mounts; 8. Cooling system mount; 9. Tunnel cover; 10. Panel; 11, 19. Tow hooks; 12. Leading wheel mount support; 13. Propeller-shaft tunnel; 15. Sidewall panel; 16. Spring support; 17. Differential mount; 18. Driveshaft housing; 20. Gearbox mount; 21. Intermediate drive mount.

Schematics Gallery

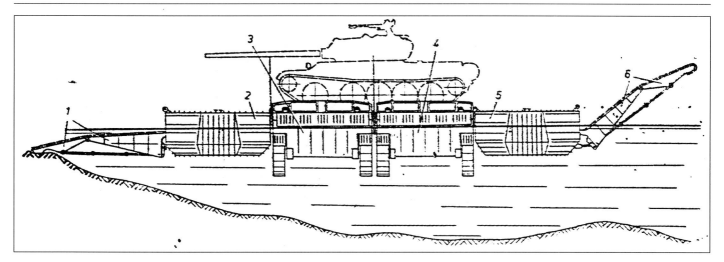

GSP-55 with medium tank: 1. Right half-ferry ramps; 2. Right half-ferry pontoon; 3. Right half-ferry body; 4. Left half-ferry body; 5. Left half-ferry pontoon; 6. Left half-ferry ramps.

Towing a stuck tank, using four pulleys and a towing tractor. The tractor is an AT-S medium artillery tractor, the tank is a T-34-85 medium tank.

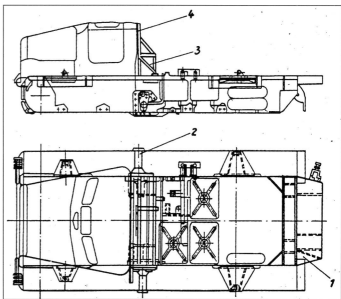

BAT-M dozer (based on the AT-T 405mu), design changes to basic 405 mu vehicle: 1. Wall reinforcement for crane structure; 2. Rack for equalizer; 3. Racks for battery boxes and hydraulic oil tank; 4. Brace for extender of crane boom.

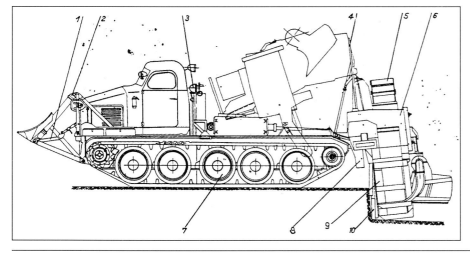

MDK-2M excavator, based on the AT-T 409mu (in working position): 1. Plow blade; 2. Hydraulic cylinder; 3. Spotlight; 4. Deflector panel; 5. Ejector apparatus; 6. Upper frame; 7. Basic vehicle; 8. Swinging frame; 9. Covering; 10. Excavator.

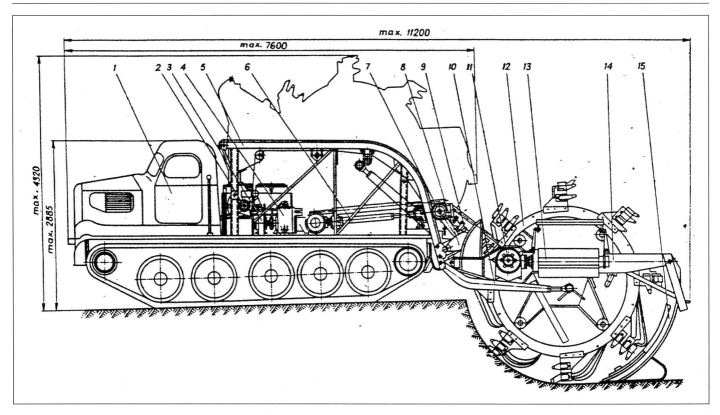

BTM-3 excavator (based on the AT-T 409u) in working position: 1. Cab; 2. Winch for digger; 3. Vertical control; 4. Divider control; 5. Frame of lifting apparatus; 6. Large link; 7. Drive extension; 8. Sliding piece; 9. Small link; 10. Extender; 11. Left side drive; 12. Excavator wheel; 13. Ejector; 14. Wheel frame; 15. Ground drag.

Below: MT-LB multipurpose towing and transport vehicle: 1. Engine; 2. Injector pump; 3. Injector pump regulator; 4. Air filter; 5. Starter; 6. Heater; 7. Upper driveshaft bearing; 8. Road wheel mount; 9. Transmission and its oil tank; 10. Main clutch disengager; 11. Intermediate gears; 12. Battery bracket; 13. Main driveshaft mount; 14. Shock absorber; 15. Shaft bearing and foot-brake mount; 16. Brake camshaft bearing; 17. Clutch drum mount; 18. Gearbox; 19. Lateral power train; 20. Steering control; 21. Clutch pedal mount; 22. Planetary steering drive belts; 24. Steering arm mount; 25. Steering shafts; 26. Clutch pedal shaft; 33. Generator mount; 36. Mounts; 37. Leading-wheel spindle mount; 38. Leading-wheel bearing; 39. Leading-wheel bracket; 40. Leading-wheel spindle; 41. Trailer hitch; 43. Ventilator drive belt; 44. Generator tension regulator mount; 45. Air-compressor tension belt; 46. Ventilator drive belt; 47. Cooler pump mount; 48. Ventilator drive; 49. Fuel injection system; 50. Fuel tank filler cap cover; 51. Side hatch bearing; 52. Side hatch latches; 53. Swivel seat; 54. Left steering control rod mount; 55. Shock-absorber attachment.

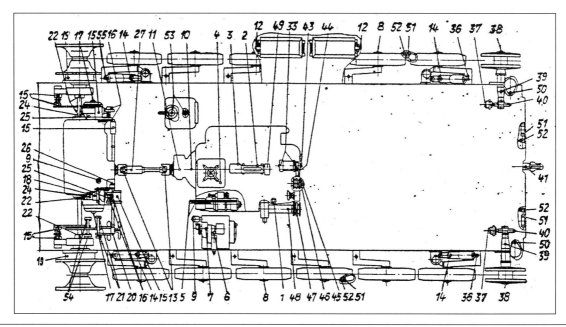

Schematics Gallery

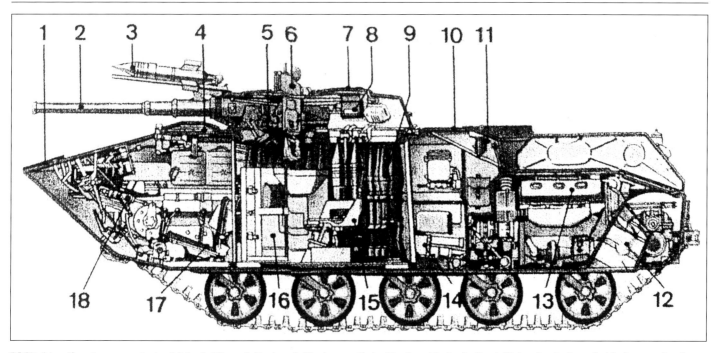

BMD-1 landing-troop combat vehicle: 1. Wave deflector; 2. Tank gun; 3. Antitank guided missile; 4. Driver's window; 5. Aiming mechanisms; 6. Targeting device; 7. Aiming gunner's window; 8. Observation device; 9. Supply of ammunition; 10. Paratroop hatches; 11. Observation device; 12. Water jet drive; 13. Engine; 14. Pneumatic lifter; 15. Aiming gunner's seat; 16. Cartridge box; 17. Driver's seat; 18. Track tension apparatus

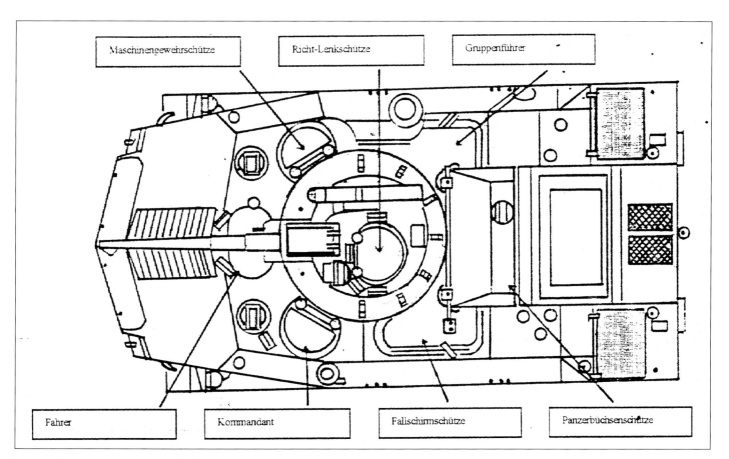

BMD-1P landing-troop combat vehicle.

183

Bibliography

Andronikow, I. G., & Mostowenko, W. D., Die roten Panzer, Munich 1963.
Antonow, A. S., Der Panzer, Berlin 1959.
Babadschanjan, A. H., Panzer und Panzertruppen, Berlin 1983.
Berchert, G., Erhart, K., Modrach, S., Otto, M., Kleine Panzerkunde, Berlin 1967.
Bonds, R., Modern Tanks and Fighting Vehicles, London 1980.
———, The Illustrated Directory of Modern Weapons, London 1985.
Damczuk, M., Czolg T-80, Warsaw 1994.
Daniel, B., Militärisches Nachrichtenwesen, Berlin 1987.
Erhard, K., Schützenpanzer, Berlin 1982.
Flotho, H., Entwicklung der sowjetischen Kampfpanzer nach Wien, Frankfurt am Main 1991.
Foss, C. F., Jane's World Armoured Fighting Vehicles, London 1978.
——— (ed.), Artillery of the World, Shepperton 1981.
——— (ed.), Jane's Armour and Artillery 1982-1983, 1988-1989, 1992-1993, London 1983, 1988, 1993.
Förster, G, & Paulus, N., Abriss der Geschichte der Panzerwaffe, Berlin 1977.
Geuckler, A., Der russische Kampfpanzer T-80U, Frankfurt am Main 1996.
Gruszczynski, J., & Rybak, E. F., Patriot kontra Scud, Warsaw 1996.
Hecker, H., Staat zwischen Revolution und Reform—Die innere Entwicklung der Sowjetunion 1922-1990, Stuttgart 1991.
Knyaskov, V., Medium Tank T-62, Moscow 1981.
Kopenhagen, W., Die mittleren Panzer T-54/55, Berlin 1981.
Korolkow, N., Panzer im Gefecht, Berlin 1962.
Krakau, A., Die sowjetische Präferenz für konventionelle Kriegführung in Europa, Frankfurt am Main 1988.
Kraynyukov, K. V., Fifty Years of Soviet Armed Forces, Moscow 1967.
Macksey, K., Tank, Enfield 1976.
Magnuski, J., "Stalin," Postrach Zachodu, Warsaw 1995.
———, Ciezki Czolg KW, Warsaw 1997.
Mayer, E., Die Panzerabwehrlenkflugkörper der Sowjetarmee, Frankfurt am Main 1991.
Messenger, C., Tanks, Barcelona 1984.
———, Tanks and other Armoured Vehicles, London 1993.
Michulec, R., Arsenal for Aggression—Armored Vehicles of the Warsaw Pact, Hong Kong 1994.
Milsom, J., Die russischen Panzer, Stuttgart 1974.
Mostovenko, W. D., Panzer gestern und heute, Berlin 1961.
Romanov, A., Bronetankovaya Technika, Moscow 1994.

Samsonow, A. M., Geschichte der UdSSR, Berlin 1977.
Schröder, S., Vom Säbel zur Rakete, Berlin 1981.
Senger & Etterlin, F. M. v., Taschenbuch der Panzer 1943-1954, Munich 1954.
———, Kampfpanzer 1916-1966, Munich 1966.
———, Der sowjetische mittlere Kampfpanzer T-34 bis T-62, Munich 1969.
———, Taschenbuch der Panzer, Koblenz 1990.
Skatschko, P. G., Tanki I Tankoviye voiska, Moscow 1980.
Skorobogatkin, K. F., Die Streitkräfte der UdSSR, Berlin 1974.
Soukup, M, Tanky 2.-3. dil, Prague 1995.
Spielberger, W. J., Siegert, J., & Hanske, H., Die Kampfpanzer der NVA, Stuttgart 1996.
Uzycki, D, Bergier, T., & Sobala, St., Wspolczesne Gasienicowe Wozy Bojowe, Warsaw 1996.
Zaloga, S. J., Soviet Main Battle Tank T-72, Hong Kong 1989.
———, T-64 and T-80, Hong Kong 1992.
———, BMP - Infantry Combat Vehicles, Hong Kong 1990.
———, Soviet Wheeled Army Vehicles, Hong Kong 1990.
——— & Magnuski, J., Soviet Mechanized Firepowers 1941-1945, London 1989.
——— & Sarson, P., T-72 Main Battle Tank 1974-1993, London 1993.

Armadni Technicky Obrazkovy Mesicnik, CSSR/Czech Republic.
Armeerundschau, East Germany, to 1990.
Defence, Great Britain.
Field Artillery Journal, USA.
Militaria - Magazyn historyczno-modelarski, Poland.
Militärtechnik, East Germany, to 1990.
Militärtechnische Hefte, East Germany, to 1990.
Motorkalender, East Germany, to 1990.
Soldat & Technik, Federal Republic of Germany
Technika I woorushenije, USSR/Russia
Volksarmee, East Germany, to 1990.
Woennye znaniya, USSR/Russia.
Znamenosets, USSR/Russia

Photos, Sketches and Drawings
Militärhistorisches Museum der Bundeswehr in Dresden (16); Author (319); Andreas Gryscheck (22), Gert Herr (34), Hans-Jürgen Janaczeck (3), Robert M. Jurga (5), Werner Klotzsche (2), Klaus Koch (132), Hans-Jürgen Riedl (4), Gerhard Thiede (38).

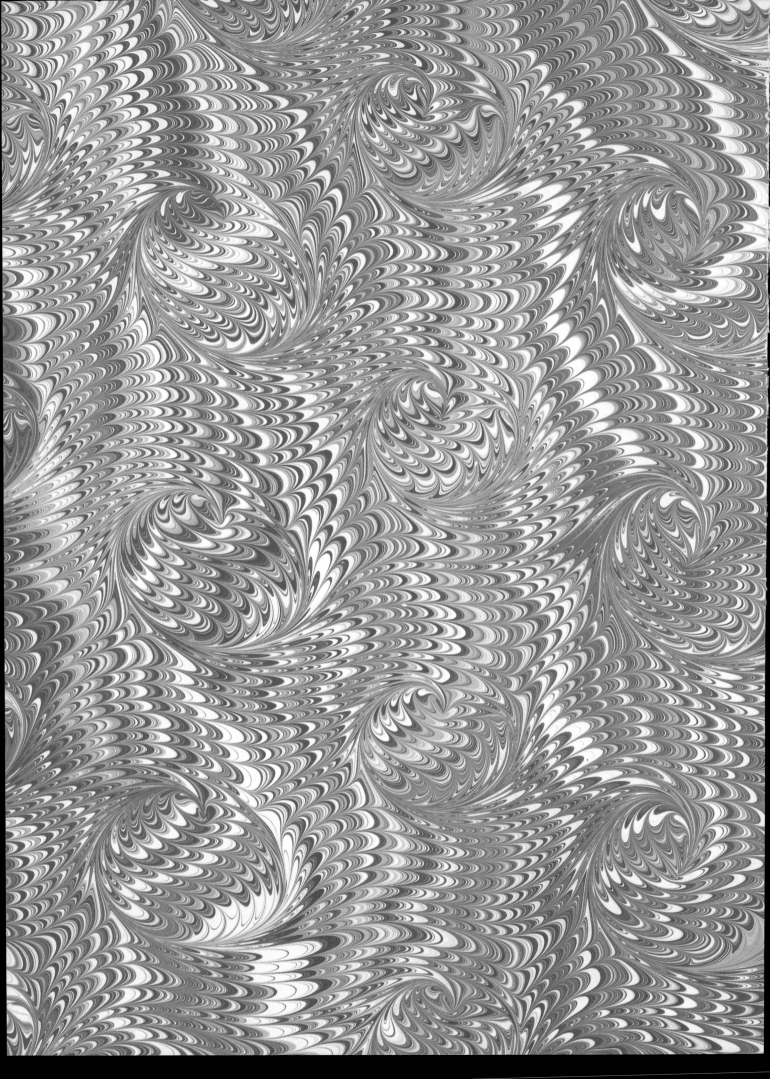